ELLIOT CONNOR

Human Nature

How to be a Better Animal

eco
ELLIOT**CONNOR**

First edition

ISBN: 978-0-6450861-1-9

Illustration by Mother Nature

This book was professionally typeset on Reedsy.
Find out more at reedsy.com

It is even harder for the average ape to believe that he has descended from man.

Henry Mencken

Contents

Preface

How does an elephant scratch an itch it can't reach? One male in Thailand found the perfect solution when he caught a car and mounted it, scraping his penis for one blissful minute against the side window. The terrified driver got away with a few hundred dollars worth of damages and the elephant returned to its daily routine having relieved its metre-long member. All up, I'd consider that situation a win-win.

So common is the feeling of an un-itchable itch that science has given it an official term: acnestis. That's the spot between your shoulder blades, between a Bactrian camel's humps, or- apparently- between a bull elephant's loins. The US patent office lists over 40 entries for back-scratchers and Milwaukee Zoo adapted a city street sweeper to service its great grey beasts. All this I mention to justify my writing this book.

For the past two years, I've had an itch, though not of the elephantine kind. This one's more of a nagging, tickling feeling at the back of my mind, telling me everything is not right with our human self-image. As you'll soon learn, it was bees and not the Greeks that invented democracy. Trees use a kind of internet called the wood wide web, slime mould can map out Britain's major highways, and dolphins have learnt to use underwater telephones. Yet we consider ourselves superior?

So I set myself a mammoth trio of tasks for this book: to define 'human,' reframe 'intelligence,' and if not divulge the meaning of life then at least describe the meaning other lifeforms give to our existence. If you can tell a hawk from a hyena or a horse from a hippo, then you know enough to follow. If not, take a look at the Animal Cheat Sheet in the appendix for handy hints and humorous asides as we progress.

My only ask is that you keep an open mind. As Einstein said: "We cannot solve our problems with the same thinking that created them." Consider this book's contents as a new filter for seeing the world, like a mantis shrimp with its inbuilt polarizing sunglasses. You'll have to struggle through the laboured similes, but in the end, you may end up with a few more lightbulbs in your brain.

And so what better way to start than with a test? You've got five minutes to answer these 12 quick questions. No cheating, no Google searches, and if you get 100% you needn't read this book at all. Your time starts now...

1.) What percentage of forest cover has been lost in the Brazilian Amazon from deforestation over the past 50 years?

a.) 20%
b.) 50%
c.) 80%
d.) Dunno

2.) Approximately how many species are we estimated to share the Earth with?

3.) How many of Earth's species has science currently described?

a.) Less than 20%

b.) Between 50 and 80%

c.) Over 80%

d.) How would I know?

4.) How has the number of recorded poaching incidents of rhinos in South Africa changed over the past 5 years?

a.) More than doubled

b.) Remained the same

c.) Less than halved

d.) No clue whatsoever

5.) To which animal group do whales belong?

a.) Fish

b.) Mammals

c.) Reptiles

d.) Hmmm....

6.) Biodiversity loss is likely to impact progress towards approximately what proportion of the Sustainable Development Goals?

a.) 40%

b.) 60%

c.) 80%

d.) ?????

7.) Since 1970, the abundance of animals in the wild has...

a.) Increased by 20%

b.) Decreased by 20%

c.) Decreased by 60%

d.) Not passed my mind

8.) What percentage of the world's economy is derived from ecosystem services?

a.) 4%

b.) 14%

c.) 40%

d.) Huh?

9.) If no major action is taken, how many degrees warmer will the Earth be in 2040 compared to pre-industrial levels?

a.) 1.5 degrees Celsius

b.) 2 degrees Celsius

c.) 3 degrees Celsius

d.) Gimme a clue!!

10.) Do richer or poorer countries have more threatened animals?

a.) Richer

b.) Poorer

c.) About the same

d.) Phone a friend

11.) Rank the following threats in order of severity for ecosystems, 1 being the greatest threat and 5 being the least.

Pollution
Climate change
Invasive species
Exploitation of organisms
Changes in land/ sea use

12.) Provide one reason why humans might be considered as separate from animals?

You'll find the answer key at the end of this preface. How did you go? 1000 people across 75 countries answered these questions for me, just as you did. We'll explore the results in chapter 2, but to ruin the suspense I'll give you a sneak preview now. Taking only the multiple-choice questions into account and removing the quality control question 5, then respondents scored on average 2 out of 8. That means a chimpanzee or a group of pigeons guessing at random could have matched them. No-one scored 100% on these questions– even the Nigerian chap who cheated by retaking the exam!

Instead of starting up a school for gifted primates, I've decided to write this book to raise our ecological IQ. Assuming you didn't get 100%, you're enrolled in this crash-course lesson. My hope is that any educated fool can understand, but I'll be asking questions as well as answering. Keep in mind that curved-ball question twelve, as we'll be chipping away at it throughout. So without any further ado, let's make a start!

Answers: 1.) a 2.) 8 million– 1 trillion 3.) a 4.) c 5.) b 6.) c 7.) c 8.) c 9.) b 10.) c 11.) e,d,b,a,c in descending order. 12.) wait and see...

1

The Big Picture

They say that dying men see their life flash before their eyes. For me, it was rather the opposite. The epiphanic moment was distinctly long-winded, my own life replaced with all life on Earth- oh... and I wasn't dying! Tucked away, shivering violently, in a snow-laden French castle, my time was divided largely between fantasizing about warm fires and caring for two dozen rough-looking birds of prey. In midwinter, no right-minded creature would be out and about of its own accord, so these were mostly old rescues- long-term care cases with injuries that would take months to heal.

In my spare time, I was both playing nanny to hordes of hedgehogs and thinking. A lot. All around me was a small menagerie of creatures placed in rehab from hunting wounds and car collisions. Meanwhile, in the news I was seeing mass fish die-offs, bat populations decimated, and frog species fighting for survival against a deadly viral disease. In every case, humans were responsible, yet they were also the people fighting against it like a Hollywood psychopath with a split personality.

I was fifteen years old, and scared. It was clear which side of our human nature had the upper hand. I felt I was witnessing some sort of mass genocide, with most other people oblivious. I started taking long walks through the countryside, all the while wondering what could be the reason for conservation's failure. It has taken me some time, but I'm on the brink of an answer. The problem is simpler than you might think.

Plainly put, we're thrifty troublemakers with big egos. Yes- that's right. And no, your book isn't missing 500 pages in- between. Check the numbering if you don't believe me. What I'm saying is that human beings are designed to be good at changing the world. It's in our genes, and like any teenager testing their limits, we've started overstepping ourselves and quickly run into issues. To get back on course, we need to listen to the parent figure of mother nature, with her 4 billion years of experience. That means embracing our heritage.

Evolution means we share over 95% of our DNA with chimps, 88% with mice, 85% with cows, 84% with dogs, 73% with zebrafish, 70% with sea sponges, 69% with platypuses, 65% with chickens, 47% with fruitflies, 44% with honeybees, 38% with roundworms, 24% with grapes and roughly 18% with bread yeast. In other words, we're not so different from animals. If you go back 400 million years in your family tree, you'll probably find that your $(great)^{76 \, million}$ granny looks awfully much like a fish. Because we evolved from fishes. That's important.

It means that our view of evolution in the human context is flawed. Every species alive on this planet today diverged from the exact same ancestor a few billion years ago. Every human, camel, fish, ant and elephant started at this one point. Every one of them has made it here in the race for survival, thus proving their lines equally well adapted to survive. Evolution doesn't favour the most 'intelligent,' and humans are getting less intelligent anyway (more on that to come). We're not all that different from other animals, and we're certainly no better. Human Nature (note the capitals) simply means accepting this. But of course, you haven't yet. So let's start at the beginning.

The human species, Homo sapiens, evolved roughly 195,000 years ago. I say 'roughly' because our pioneering ancestors did not, unfortunately, hold a great big party to mark their genesis. Nor did they bury a time capsule (aside from several preserved footprints and skulls) to let future generations know of this momentous occasion. The process itself took upwards of a thousand years, and the result was far less striking a difference than between, say, you and Genghis Khan.

The difference between two species is, in fact, rather more arbitrary than we like to make out. Ask your average biologist what a species is, and chances are they'll come back with the common knowledge answer: "A group of animals in which individuals of opposite sexes can interbreed." This is what we learn in science at school, in media and popular culture in all its many forms. Just think of the 50-odd women James Bond has hooked up with and it all makes perfect sense!

Sadly, simplicity doesn't always equate to truth, and this basic grouping carries with it a selection of shortcomings. What exactly does 'opposite' mean for a microbe with seven sexes, or in fungi boasting over 20,000? And what about Komodo dragons, in which females can clone themselves without sexual intercourse? Or hermaphrodite slugs for that matter too? Not to mention all the many hybrids that exist: from wholphins to ligers, narlugas to zonkeys to grolar bears. You can imagine the confusion.

3

The Oxford English Dictionary (OED) makes a valiant, if futile, attempt to provide a solid answer. A species, it says, is "A group or class of animals or plants having certain common and permanent characteristics which clearly distinguish it from other groups." Not much to argue with there. It's so vague it could fit around pretty much anything, and could just as easily divide men from women, or Russians from Chinese. It's also as good a description as we've got.

A species is just shorthand for where scientists choose to draw the lines between similar animals. These are clusterings on a spectrum of animal-ness. That's about as scientific as it gets. The OED actually includes an afternote: "The exact definition of a species, and the criteria by which species are to be distinguished have been the subject of much discussion." Hmph.

Back to the story: for millennia, humans lived and (more often) died in harmony with nature. Large predators and internal disputes kept our numbers sustainably low, whilst lacking heavy artillery and nuclear weapons meant we couldn't kill ourselves off completely. All was well enough with the world. 10,000 years ago (give or take), we started farming, and that meant grouping together in large communities for the first time. Civilizations rose and fell until along came the industrial revolution in 1800 that blew everything out of proportion. In 200 years, the human population grew from 1 billion to 7.7 in an unprecedented exponential rise. The rest, as they say, is history.

You probably know this. It's your own story, after all. How you came to be here, right now, reading this incredible book, wishing it would only get a move on. I've got a list of nature puns in front of me that I'm dying to use, so let's stop beating around the bush and get straight to the point.

Our world is changing faster than it ever has before. We're encountering new opportunities and challenges daily- a sort of evolution on steroids. With all this rapid progress, other things are thrown aside, like caring for our environment. Remember that humans are the new kids on the block, adolescents discovering their own strength. We're so wrapped up in our looks, exam grades and secret crushes that we don't think twice about making a mess of our home. The pigsty bedroom has been replicated in Earth's ecosystems.

Let's pause for a 'What if' moment- because I sense the more sceptical among you are wondering why it's so important we conserve nature in the first instance. It goes like this- What if I could snap my fingers and make all non-human life disappear? Would we notice? Do humans even need nature at all? What would it matter?

Take, for example, an evening scene where you're juggling pots and pans in the kitchen. You may have potted plants indoors, in which case these would be gone in a heartbeat. Fido would be cut off mid-bark, and the ambience of bird and insect calls we've learnt to tune out would be painfully loud in its absence. Your garden would be stripped as all the plants, weeds and grass were removed, and there'd be some serious issues with soil caving in as insects and micro-organisms departed. Those would be some of the obvious, immediate effects.

As it happens, very few people have had a serious look at this question. They say they've got more important things to discuss. So I'll do my best to set out the consequences. Firstly, we'd have no more food production. Crops and cows are plants and animals, so they've all gone "Poof!" and disappeared. All that remains is what we've shelved or stored away, which being (mostly) dead would have been saved from the finger snap. Headlines in 2012 stated that food stocks globally were at a record low- sufficient only to last global consumption 74 days. The situation has since improved... but only slightly. Combine that with discrepancies in access to resources, and it's safe to assume we'll have billions starving in the space of a month (alongside total anarchy).

The next question that arises is what happens to the air we breathe? Algae and bacteria in our oceans produce most of our oxygen, assisted by plants on land. With all of these sources wiped out, it's easy to see why we might run into difficulties. The good news is, we've got some pretty large stores of the gas stuffed into our atmosphere. This 1200 trillion tons is enough to get us by in the short term. And by short term, I mean 100,000 years under current human consumption. The challenge would be getting it down to where we're all living and breathing.

Other issues aren't so considerate. Aside from food short-ages, our digestive systems themselves would be having a hard time of it. Deprived of the bacteria needed to break down nutrients, chances are that most of us would be losing our appetite, suffering from severe constipation as we watch the end of the world. We'd have widespread landslides and desertification without plant roots to stabilize soils, topping out a cocktail of dramatic weather shifts and natural disasters. Most disconcerting is the fact that viruses might still be around to plague our weakened immune systems. Science is divided as to whether these critters can be considered as living or not. It certainly would be a quick way to find out.

Now, this might seem like a meaningless prophecy, but we're rapidly making it into a reality. Scientists are, on the whole, an unassuming bunch. So when they start adding phrases like 'insect armageddon' and 'biogeochemical asphyxiation' into the jargon, you can tell that something's up. 2020 marked a big year for biodiversity because 2/3 of all wild animals have now been lost since the moon landing. It has been dubbed a sixth mass extinction. Some species cling on, but go 'functionally extinct' once their populations drop so low that they can no longer play their role in an ecosystem. This results in ripples across all the other creatures they interact with, destabilizing food webs.

If you did your homework and read the preface, you'll have come across the Sustainable Development Goals. The SDGs, as they're known, are a set of 17 major milestones adopted by all UN countries as a framework for improvement lasting until 2030. They span everything from poverty to healthcare to education, employment and gender equality. Suffice to say, biodiversity loss hinders progress towards 80% of these. Ecosystem services like crop pollination, air and climate regulation provide for 40% of the global economy, and 80% of the needs of the poor. Thus their loss is felt most keenly by the disadvantaged, who have done the least to cause it.

Are you feeling miserable yet? Maybe. If not, you, like most people, have probably heard so much of this doom and gloom rhetoric that you frankly couldn't care less. Perhaps you've stopped reading, but somehow I doubt it. In either case, what's needed is a clear alternative: one that requires minimal effort from you and fixes the whole mess of a situation nicely- a silver bullet solution for our environment's woes. This is what the Human Nature worldview provides, and the secret is...

Woah woah woah woah woah!! Not so fast! A good raconteur never gives away a story's plot at the beginning. You'll just have to wait for the happy ending. It is coming, I promise. For now, let's continue our journey down this vein of thought and see where it ends up. We've shown that our relationship with nature could use some improvement, that humans being a species is a pretty meaningless division, and that we need nature to survive. It's time to add some more strands to this web.

There's a remarkable fluke of nature that means a pair of headphones placed in your pocket inevitably gets tangled within a minute. So imagine this: a string the length of all your DNA, equivalent to twice the diameter of the solar system. It's a fine thread, like a fishing line, and it's jumbled up alongside the DNA strings of every other living creature on Earth. I bet you'd get some pretty nasty knots that form.

The point of this long-winded analogy is to illustrate just how closely life is connected. Have you heard of the butterfly effect? The story goes: one flap of a butterfly's wings is enough to cause hurricanes on the far side of the planet. Whilst this might be a teeny weeny bit of an exaggeration, it does hold a grain of truth. No mathematician could fail to appreciate the complexity of weather prediction equations. Solved with countless data inputs from stations across entire nations, these formulas require precision to the Nth degree for accuracy in their outputs. A cloud here or there may be enough to lead the whole system astray, ruining a week's forecast and leading to a host of unnecessary beach party cancellations.

Take a historical example. In 1907, the young Adolf Hitler applied to study at the Academy of Fine Arts in Vienna. He was rejected. Twice. Told his talents were better suited to architecture, he was forced to turn to other means for sustenance. Architecture, unfortunately, wasn't something he fancied.

Still, he famously stated shortly before WW2's outbreak: "I am an artist and not a politician," expressing hopes of a peaceful retirement to nurture his passion. Personally, I think he wasn't a bad hand when it came to painting, and you can see some of his works online. If only he'd been accepted, many lives would likely have been saved. Of course, the joke now is very much on the academy. Each of his remaining works is worth tens of thousands of dollars, some auctioned off by private collectors and others locked away by the US government.

What's worrying about the butterfly effect is when we apply the same logic to life on Earth. The American aerospace engineer Edward Murphy Jr. never met a drunken hippo, but Murphy's Law stating "Anything that can go wrong, will go wrong" applies to animals and ecosystems all the same. Animals don't like following rules any more than humans do. That makes them hard to predict, and if you scale that randomness across landscapes, the results can be dramatic.

The classic example is Yellowstone's wolves. In 1872, Yellowstone National Park was formed- the first example of its kind. Except it wasn't what one might consider a sanctuary by today's standards. Government regulations encouraged hunting of all animals for the first decade of its history, and all predators thereafter. When the US military took over management in 1886, it was actually a relief for the park's animals. All hunting was banned, and predator control reserved for officials who largely neglected the task. 1916 saw the freshly formed National Parks service assume control, and an escalation of killings. 140 wolves were slaughtered by rangers in the space of a decade, right up to the final two pups in 1926.

After that, there were no more wolves in Yellowstone. They

were extinct within the park, with scattered remnant populations roaming the surrounding countryside. And that was that.
The populations of elk they had once hunted went through
the roof, and the entire ecosystem suffered. The woody plants
they browsed on were overgrazed, unstable soils eroded, and
when deer culling ceased in the 1960s, the effect was only
exacerbated. Then things started to change.

1966 saw the Endangered Species Preservation Act come into
force. The grey wolf was amongst the first species to be listed
in the US as <u>Endangered</u>. Years passed, laws changed at their
own sweet pace, and governments came and went. Momentum
grew, plans were suggested and discarded, but finally in 1995
wolves were reintroduced to the park.

From an initial pack of 14, the wolves wasted no time in
reclaiming lost ground. In the space of three years, the
population grew to 100. In a decade, elk populations had
halved- each wolf taking down 22 of them on average each
year. The forests regrew, birds returned, grizzly bears and
foxes rebounded. Beavers returned en masse and the dams
they built created new habitats. Otters, muskrats, fish and
amphibians all showed up in greater numbers thanks to the
slower, broader rivers.

In short, the removal and reintroduction of wolves to Yellowstone completely altered the ecosystem. That's why they're
known as a 'keystone species'- the beer coaster beneath a stack
of cards. Remove them and everything else collapses: few
could have predicted the full extent of the consequences their
absence brought. It taught us a lesson, and great caution was
taken in projecting the effects of their reintroduction before
it was performed. Once we appreciate nature as being 'bigger
than us,' we're better able to accept these sorts of processes

and take care in our meddling.

"Life finds a way," as the inimitable Dr. Ian Malcolm once said. There's a certain beauty in the resilience of natural systems and their remarkable ability to rebound. But this elasticity has a limit, like a ball of silly putty you might have played with as a kid. Given time and space, it will spread out to fill the area available. Yet if you squeeze the ball, you'll find that it readily deforms; stretch it too fast and it snaps. 99% of all species that have ever existed are now extinct- gone, forever. So here's a twist: what if we could bring some species back from the void? What then would become of nature?

It's an interesting notion, isn't it? And one that is far from impossible given a decade or two of scientific advancement. In the year 2000, just as everyone was celebrating the turn of the millennium, the Pyrenean ibex went silently extinct. This mountain goat had been hunted relentlessly for the past few centuries. By the time conservation action kicked in around 1990, only a dozen remained. Slowly the numbers dwindled until a single female named Celia was the last of her kind. She was caught and radio-collared to protect her from poaching, but beyond that little could be done. It was a case of too little too late. Nine months later a monotone from the tracking signal indicated that she had passed away- crushed beneath a fallen tree trunk. Game over for the ibex.

Then in 2003, a team of scientists managed to turn back time... momentarily. Using parts from frozen ibex cells injected into goat eggs cleaned of DNA, they successfully produced 57 pseudo-ibex eggs with which to revive the species. These were implanted into surrogate goat mothers for gestation. And they waited. Only 7 of the goats became pregnant and 6 of them suffered miscarriages, but after an exhausting effort made by one brave mother, Celia version 2 was born. Having been cut out via Caesarean section and weighing a mere 5 pounds, the young ibex clone was dead within ten minutes. A malformed lung had laboured her breathing, starving her of oxygen. So there's no happy ending yet.

Still, we're unbelievably close to making de-extinction a reality. And whilst I'm first in line for recreational time-travel, it does beg the question: how on Earth does this unparalleled power fit in with conservation? Short answer- it doesn't. Because our theories and philosophies were out of date before they'd been conceived. We shouldn't be trying to restore the world to some prior state. Change is natural, and how far back do you have to look anyway before 'peak nature' has been reached? Instead of returning nature to a half-forgotten past before humans had screwed everything up, we need to bring nature forward in a way that improves outcomes for animals and for people. De-extinction could have a place in this model, but it does have its drawbacks. Here's my take:

De-extinction cannot exist for de-extinction's sake. Jurassic Park, fictional world that it is, can still teach us something. Animals don't follow rules, remember? Think of Yellowstone's wolves, the butterfly effect and our old friend Hitler. Chaos isn't pretty, even if there are no Tyrannosaurs to go around munching people. There needs to be a very good reason for intervention. Perhaps a T-rex could stabilize an ecosystem, or help to deter rhino poaching (I imagine it would!)?

Luckily, with our current scientific understanding, it seems we're limited only to species that have gone extinct in the past few tens of thousands of years. Beyond that, the DNA is too fragmented for its life code to be recovered. That means no T-rex and no Jurassic Park. It also means that most of our target species are those we as humans have directly contributed to the extinction of, giving us a moral obligation to act. It's not so much a matter of 'playing God' as pressing undo on our previous Godlike actions that got us here in the first place.

That's where things get complicated. Is it ethical to bring back an animal merely to represent its kind? To keep it confined in some artificial habitat facility for us humans to gawk at? You must recognize that when bringing back, say, a mammoth, the places it used to call home will have changed beyond recognition. Where does it go? In a single century since the Tasmanian tiger, or thylacine, went extinct, its environment has changed dramatically. Would it survive if it were reintroduced? And what about all the other species that spent generations not having to watch out for this predatory threat? It's hardly fair to chuck them in the lion's den by bringing large carnivores to their doorstep.

There are all sorts of arguments why it might be a great idea. By recreating extinct lifeforms we're better able to study their behaviour, assisting ecologists in understanding extant species and possibly leading to medicinal breakthroughs. Then there's the tourist revenue, as people flock to these sights to catch a glimpse of such impressive creatures. Just as ecotourism funds can be directed into conservation, so these new attractions can be made into life-saving cash flows for the resource-starved field of environmental management. Both are a double-edged sword due to moral concerns, but with the potential to cause far more good than harm.

Pleistocene Park in Siberia is the closest yet we've come to this sort of vision. The brainchild of two Russian scientists, it forms a large enclosed area into which a suite of animals have been introduced. The goal is to recreate the conditions of 10,000 years ago when mammoths roamed. The results are both promising and surprising. Having placed everything from bison to moose, yak, wild horses, muskoxen and elk on the property, the environment has changed completely from barren tundra into rich grassland. Permafrost thawing during summer has dropped steeply, leading to lower greenhouse gas emissions from the land. After all, it was human hunting that drove this ecosystem into decline, so why not restore it? And if that helps tackle climate change at the same time, well then that can be considered a bonus. The missing piece in this puzzle is, of course, the woolly mammoth. Who knows when that will return?

Perhaps the strongest argument against these efforts is that the millions of dollars of funding currently flowing into R&D for well-stocked laboratories might be better employed in conserving the most critical cases amongst our current biodiversity. Perhaps having the ability to resurrect lost species would mean even less action to protect natural spaces? There are still dozens of as-yet undescribed species going extinct daily, which de-extinction couldn't save. So the final word in the debate would have to go towards use cases in conservation practice as we know it.

Take the two species of gastric brooding frog that inhabited Australia in the late 20th century. Almost as soon as science discovered them, they disappeared, and have never been seen since. Both share a unique biology that only increases their scientific worth. Plus having disappeared in the mid-1980s, the genetic material recovered from them is remarkably well preserved. They could be brought back into existence well within a decade if we so desired. Their ecosystems are accustomed to their presence, which supports a plethora of other species as well. Then there are the 200 mammal species classed as 'Critically Endangered,' many of which are subject to unsuccessful breeding programmes in zoos. Cloning these remaining individuals would make a world of sense, and relies entirely on de-extinction technology. Bringing back the recently deceased (animals, not humans) is even better, preventing inbreeding in these small remaining groups of animals.

A lot of the reticence to this and other ideas lies in tradition. Because whilst conservation in the modern sense has been around for just over 50 years, its roots go back much further. Did I mention I like time travel? Well, buckle up because you're in for a ride! Introducing "Pigeons and People: A Bird's Abbreviated History."

Pigeons and doves (Columbidae for the more scientifically inclined) arose long before humans, but once we showed up on the scene they were quick to spot an opportunity. Self-domesticating long before recorded history, they worked their way into our hearts with their intelligence and tenderness. Bronze Age paintings show the birds alongside the Mother Goddess in ancient Mesopotamia. If there's one sure way to safeguard your survival, it's to make yourself sacred. Most of the time.

With the rise of the Greek empire, these same divine symbols were adopted for Aphrodite- the goddess of love. Doves, therefore, found themselves on the wrong side of immortality: being served up as offerings. Why they thought beautiful women have a taste for dead doves is beyond me, but for the birds themselves, this was far from good news. A few lucky ones found for themselves rather more illustrious careers. Some were used to announce the victor of each Olympic games, others to carry important messages in times of war between the city-states. So much for the dove of peace.

Aristotle began the practice of describing nature scientifically and covered pigeons in considerable detail. He described the act of kissing that precedes mating and their varying methods of evading birds of prey according to each one's hunting style. The birds were easy to observe, relatively abundant, and widely used for all manner of tasks, hence our understanding of them was good. In other situations, as with migrating birds, readers were not so fortunate. Aristotle convinced himself that many species simply transform without moving at all. Thus the robin that appeared in greater numbers over winter was, in fact, another bird- the redstart- adopting a new guise. Meanwhile swallows, he claimed, chose to hibernate at the bottom of lakes overwinter. You can't be right about everything!

Roman historian Pliny the Elder elaborated upon this trend in his 37-volume epic Naturalis Historiae. This he humbly described as covering 'life.' Then again, why narrow your ambitions when you've got servants to carry you, write for you, bathe you, feed you and so on? As the origin of natural history, the work left much to be desired. And of course, pigeons again featured prominently thanks to their central place in everyday life. Thus, descriptions of their drinking were accompanied by those of marriage and domestic life, a strange affinity of theirs to kestrels matched by details of a prayer to venus they uttered during courtship. With such an imagination it's not hard to see how readers fell for his works, and the enduring myths they created that persisted for centuries.

Despite an increase in demand for candles, the dark ages were... well, relatively dark for zoology. The Norman invasion of Britain saw the rough Germanic language blended with French in the birth of modern English, and so we arrived at the current confusion of terminology. The prior Germanic "dove" was joined by the French "pigeon." The latter, locals reportedly reserved for mundane uses in cooking, sport and breeding as an insult to the invaders. Today the terms are synonymous. Sometimes doves are smaller than pigeons, sometimes the opposite. They really are the most pointless dual terms. The moral being that communication is difficult, and languages are barriers. Anyone who insults scientific naming deserves Damnatio ad bestias. I'll leave you to work out what that means.

In 1662, one of the most famous pigeons- the dodo- went extinct, a mere 60 years since its first sighting by Dutch sailors. Hunting, invasive species and habitat loss all played a role in its rapid demise, but the root of the matter is that no-one at the time cared enough to even describe it scientifically. Much less conserve it. Carl Linnaeus appeared a century later and invented taxonomy for categorizing animals, alongside the index card.

From here things pick up pace. Gentlemen scientists start creating cabinets of curiosities that morph into the first museums. The Natural History Museum takes form from the collections of a certain Sir Hans Sloane, and despite an early habit of staff to periodically cremate thousands of specimens, this 'cathedral of nature' has miraculously retained some 80 million items. Amongst these are 300-odd pigeon species. Sir Richard Owen took over as director in 1856, and besides coining the term 'Dinosaur' and keeping a rhinoceros carcass in his hallway, spent much of his time arguing vehemently against Darwin's theory of evolution. For the uninitiated: scientists can sometimes find consensus. Naturalists NEVER agree. David

Attenborough was also born around this time.[citation needed]

Nowadays, Charles Darwin is almost a household name- the great naturalist who ate most animals he encountered and completed half a degree in medicine despite a phobia of blood. Somehow, most people still know him for describing evolution. Presenting his theory in the catchily named <u>On the Origin of Species by Means of Natural Selection, or the Preservation of Favoured Races in the Struggle for Life</u>, it's a miracle the book did as well as it did. Goodness knows how that fitted on the cover. Darwin drew upon both his time sailing the world and his hobby as a pigeon fancier in constructing his ideas on the survival of the fittest. Pigeons were the lab rats of the day and formed a large part of his proof. He also let his kids doodle over the book's manuscript, which just goes to show that people who spend their time dissecting animals aren't as creepy as you'd think.

Pigeons in Victorian England saw an uptick in popularity as the queen herself took a fancy to the Jacobin breed. She was, however, one of many notable persons to show fondness towards the birds. Pablo Picasso named his daughter "Paloma," meaning pigeon, and the 16th century Mughal ruler Akbar the Great kept 10,000 of them in his private collection. Even Nikola Tesla took a strong fancy to the birds, rescuing injured ones and bringing them back to his hotel room. One favourite he described as being the closest he would ever come to a wife. That's what I call love at first flight!

With the emergence of pigeon racing as a sport, prices for well-bred birds skyrocketed, and nowadays an individual might sell for upwards of $300,000. Although their average speed is about 100kph, they've been shown to fly faster through polluted air. Turns out good old Queen Vic's London smog would have made some record-breaking birds.

Recent history saw wars steal the spotlight, and doves served with distinction in the Franco-Prussian war as well as in Palestine's War of Independence, before being banned by the Taliban for fear of subterfuge. One pigeon named 'Cher Ami' was awarded the French Croix de Guerre award after WW1, having delivered 12 messages amidst a bullet wound and the loss of one leg. Likewise, 32 pigeons bearing names such as G.I. Joe, Mary of Exeter, Paddy, Commando and William of Orange all were honoured with the Dickin Medal for service after World War 2. Dare I say they deserved it after being air-dropped repeatedly over enemy territory. Worse would have been conscription into the US army's Project Pigeon for the creation of guided dove missiles!

Pigeons formed the first regular airmail service in New Zealand, dubbed with the rather clunky name of 'Pigeongram.' They were used to deliver medicines to hospitals, and to take some of the very first aerial photographs. All in all, we owe a lot to these rats with wings. But it would seem things are going downhill, wouldn't it? The Romans and Greeks never locked their messenger birds in explosive metal containers. Nor did they create elaborate pigeon-houses for the express purpose of stealing and destroying their eggs. Avian contraception hadn't been invented in those days, but even so, I doubt they would have used it.

There's a certain elegance to using birds such as these in exploring the history of our connection with other creatures. Our history and theirs for the past few thousand years has run largely parallel in course. Where we've gone, they've followed, the result being that rock doves can be found in cities on every continent except Antarctica. Thanks to their familiarity, we can trace as our knowledge of natural history expanded from hearsay and Chinese whispers to systematic observation and analysis. Thus the explanation of 17th-century scientist Charles Morton that birds migrated seasonally to the moon was debunked when in 1900, numbered metal bands were placed on birds' legs for observers to record and track their movements.

It is these animals more than perhaps any others that capture our imagination. Think of what a powerful symbol the dove and olive branch makes from the Bible story of Noah's ark. Or the dodo Lewis Carroll inserted into Alice's Adventures in Wonderland as a caricature of himself. We are programmed genetically to resonate with other lifeforms because doing so helps us to learn and to survive. Now perhaps more so than ever. As we've built ever more concrete structures and high walls around ourselves, our instinctual empathy towards animals has been challenged. It is in no way diminished.

What links pigeons to saving the world? None of this long account might sound much like conservation. That's because conservation is a modern creation, starting in the 1960s with Rachel Carson's Silent Spring, the publishing of the IUCN's first Red List of threatened species, and the emergence of the World Wildlife Fund with its ubiquitous black and white panda logo. That means we have fifty years of solutions versus five hundred years of accelerated development and globalization at nature's expense. The tradition plaguing the system comes from the latter just as much as it does from the former. You can see why that might create some challenges.

So pigeons are an indicator of our perspectives towards other animals. We've been through an age of enlightenment, an age of exploration, and are struggling to shake off an age of exploitation of other living creatures. What I set out in the chapters that follow is merely my way of continuing the trend in a more positive direction, towards restoration and the re-education of humanity. So on to the happy ending...

Hang on hang on hang on. Hold your hippopotami! Not so fast. If I'd cut to chapter two there, I would have been letting you off too easy. There's one more thing that my little dove diversion shows, and that's how much better life is for humans and pigeons when people stop fighting the world and try to better understand it. Because I'll let you in on a little secret: 'survival of the dictator with the funniest facial hair' doesn't work. We've barely scratched the surface when it comes to learning about nature, and the case in point is animal intelligence. We've questioned the word 'human,' taken a new stance on nature- now it's time to blow up myths about braininess.

My favourite animal changes often, like any good parent with a favourite child. Right now, the title goes to the African elephant, after some quality time I spent up close and personal with them in the wild. If you're ever after a perfect example of a model family, look no further because this is it. They can tell how full a cornflakes packet is by smell alone, will walk 200km in a day without complaining, and give younger siblings their trunk to suck for comfort. Such is the amazing complexity of their social lives.

Ever heard of elephant graveyards? For whatever reason, elephant bones have a habit of finding themselves piled up in certain locations. No-one knows quite how this happens, but what we do know is even more interesting. Elephants will stand for days mourning and weeping over deceased members of their herd, burying them in dirt and branches. They go out of their way to help other animals including humans, described by Aristotle as "The animal that surpasses all others in wit and mind."

The reason for this harks back to what's called the social intelligence hypothesis. The theory goes: animals that operate within larger groups display more 'human' mental capacities because living in these social dynamics necessitates it. If I were to list out the 50 most intelligent animal species, chances are that 95% of them would be found in family groups or larger collectives, as opposed to a solitary existence. One of our defining characteristics as humans is our ability, through technology, to collaborate over global scales. So does that make us more intelligent than other creatures- more adept at problem-solving and communicating effectively?

Think about it. Use that oversized brain of yours and do the maths. Over the few hundred thousand years Homo sapiens has existed, only for the past few hundred years have we had the technology to move quickly and spread ideas over any large distances. Sadly evolution doesn't work that fast, and your brain is practically indistinguishable from that of a Roman soldier. Take that as you will. Yet there are changes that are noticeable irrespective of whatever brain bulk you might have locked away in that skull of yours. Because some things do change over such short spans. Like food.

Now it may come as a surprise to you when I say that Uber Eats wasn't around at the dawn of civilization. Even popcorn had to wait for Columbus' trip to America before it could emerge. So nowadays we're eating much better than most of your distant ancestors would have done. That makes a big difference for your brain. Properly nourished, you're able to devote more of your energy to doing all the big important things like philosophizing about Netflix shows and reality TV.

That big lump of jelly inside your skull demands a whopping 20% of your daily energy expenditure. How selfish.

And that's not the last of it. Healthcare and education are changing too, mostly for the better (so much for the good old days). We've gradually phased out the use of lead paints and beat back infectious disease, increased the average years of schooling, lowered smoking rates, and started having progressively smaller families. The same thing is happening across the globe, regardless of affluence, religion or ethnicity. This is wonderful news. And like all good news, there's a catch.

James Flynn is an extraordinary man, not only because of his superb facial hair (think Einstein times two) but because he has shown our IQ is on the rise. Great! So I'm smarter than my parents, and as for older generations having authority... well, we'll see. Overall, our fluid intelligence (problem-solving, reasoning etc.) has risen by 15 IQ points per generation since the 1950s, with a still-respectable 9 point rise for crystallized intelligence (stuff you've learnt). You're using a different IQ scale now than your parents might have used because these tests are continuously standardized to get an average of 100 and the right spread of scores. That, and so that you don't realize you're a genius!

Then things get really interesting. Many new studies have shown that since the 1990s, the exact opposite is happening. In certain countries, people are getting dumber: Norway, Denmark, Australia, Britain, the Netherlands, Sweden, Finland, France and German-speaking countries are amongst those that have been tested and shown this trend. The reason? We have no idea. Air pollution may be a contributor, or it's possible we've been selecting partners for non-IQ factors so long that we've caused ourselves to evolve stupidity. Either way, I think

23

it fair to say that our technology has not made us significantly smarter. Standard of living perhaps, but that's a short-term boost and then nothing. I'm sorry- Skype just doesn't cut it for picking humans as the dominant species.

Quite a pickle, isn't it? Social intelligence can't so much explain what makes us, 'us.' Technology falls short. And to cap it all off, everyone seems to be getting dumber yearly. Seeing that I've deflated your ego so dramatically, I feel honour-bound to lift it back up. Humans are smart- I'm the last person to say otherwise. And given the family groups we naturally form, there's certainly a social element involved too, just as there is for dolphins or wolves. Let's flip the image for just one minute and look at humans from a biological perspective. What defines the human animal?

Essentially, humans are opportunists. We'll eat just about anything, from plants to meat and more recently chemicals which are neither of these. Our lifestyle is flexible, and as we've spread across the globe we've adapted to the changing conditions that each new location holds. Like the raccoons that raid peoples' bins at night or the rats that live in your walls, we have shown a tremendous ability to capitalize upon chance occurrences and learn from others' faults. Passing down this knowledge through generations is the basis of culture.

What defines opportunists such as ourselves is the ability to explore new locations and modify our behaviour to fit into new surrounds. That's why humans have spread to every continent on Earth and set up large permanent settlements on all of them except for Antarctica. The Norwegian rat (that came from China) has followed us in doing so, hence can claim a similar showing. In fact, as fellow opportunists, we share a lot in common. We both like being tickled, hate eating celery, enjoy fishing, prefer new partners for sex, live in large, hierarchical groups and can commonly be found in cellars. Who'd have thought?

Jokes aside, there's one more half to the story. Besides being excellent opportunists, we show a particular capacity for changing our environment. That makes us "ecosystem engineers." Remember the beavers of Yellowstone national park? Once the wolves returned, they flourished, and in doing so completely changed the watercourse of the valleys. They created new environments for a suite of other animals. Humans do likewise, only on a much larger scale. We're creating new urban environments for ourselves, modifying landscapes to grow our food, and changing river flows for electricity and irrigation.

By moulding the world around us, we circumvent the need to adapt, saving precious time and energy. We can fire rockets or fly planes through clouds and cause rain to fall, hurricanes to change direction, or hailstorms to subside. We can launch ourselves to Mars, dehumidify its air for drinking water, create oxygen from the 96% carbon dioxide in its atmosphere, and start a colony there by 2025. We could even begin terraforming the Red Planet, using a solar sail to sublime the dry ice at its poles, thickening its atmosphere to raise the temperature and block out radiation, generating running water, rain and snow, increasing the pressure so we no longer require spacesuits and allowing more crops to be grown.

The bottom line is that humanity with its ability to morph its surrounds has done what few creatures have ever done before. We've attained a level of control of our environment no animal alive today can match- the latest in a long line of dynasties. Once it was fishes, then insects and fungi, amphibians, reptilian dinosaurs and now the birds and mammals like ourselves. We still wouldn't be here if, 2.4 billion years ago, cyanobacteria hadn't orchestrated a huge rise in Earth's oxygen levels, allowing complex life to emerge. Our society could be obliterated in an instant by an asteroid impact with mere days of warning, or by a pandemic to which no cure can be found. Despite how far we've come, we're still vulnerable, and nature acts as a safety net in moderating global disasters. The coronavirus outbreak at the start of 2020 wouldn't have come about if the illegal wildlife trade had been stopped. It's all connected.

Being in this unique position we inhabit, we are faced with both challenges and opportunities. Ever wondered why we need to grow old? How much nicer it would be to live with chimps? Even whether you need a brain to think? By taking a step back and looking at ourselves through the Human Nature worldview all of this and more can be answered. Are you ready? Then let's begin.

2

Life's a Toothbrush

I never wanted to be an astronaut, but I have always wanted to put my name on a species. I mean, there are enough of them out there, so who's to say I can't? David Attenborough has a whopping 18 named after him, and Darwin steals 285 plus a space mission. Everyone who's anyone, from Mandela to Mozart, from Bill Gates to Beyoncé, Harrison Ford, Douglas Adams, Einstein, Obama, Charlie Chaplin, Sigmund Freud, Lincoln and Lennon, Tolkein and Trump has a species for themselves. Sure some of them do better out of it than others, from the relatively glamorous birds, bees and fish to the passable wasps, beetles, spiders and the slightly insulting parasitic flatworm– but in the end, a species is a species, and I'd take an amoeba in an instant.

So what is stopping me from attaining this lasting legacy? Quite a lot, actually. Firstly I'm not famous, which throws a stone in the works. Secondly, science is slow (no offence), so whilst we've got more species than you could shake a stick at, science has to place them all under a microscope to properly describe them. Did you know that less than 20% of species have been classified? This is one of the questions from the quiz in the preface that people did most well on...

3.) How many of Earth's species has science currently described?

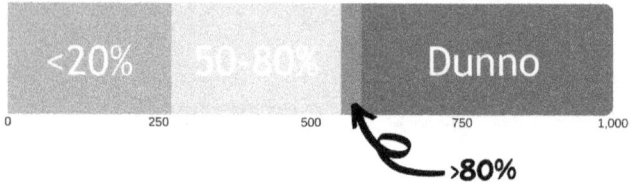

In fact, the figure could be much, much less than that, because the simple truth is: we don't have the foggiest idea how many species there are here on Earth. Consider for a moment that we can look way out into space and make some fairly good estimates of how many stars there are. Our milky way galaxy has at least 100 billion, and the universe has 1 septillion (1,000,000,000,000,000,000,000,000) give or take a few. But when it comes to knowing far simpler stats about the life right under our noses, we're clueless.

Traditionally, the accepted figure for Earth's species count has been 8.7 million or thereabouts. Three-quarters of these were said to inhabit the land, and one-quarter of them the oceans. Simple, right? 1.74 million have been thus far described, with about fifty being added every day (told you it was slow). Those are the figures the 20% was based off, and which are still some of the most widely used. They are not, however, the only ones. More recent estimates peg the number at anywhere between 2 and 10 billion, whereas some clever maths and scaling laws (ratios of big critters to small critters) saw a 2016 paper claiming it may be as high as 1 trillion. In short, no-one has a clue.

Okay- so we're not really sure how many species I might have to work with, and they're being listed... well, a little too slow for my liking. Then there's also that inconvenient truth of extinction to grapple with. Half of all species are booked in for complete eradication by the end of the century, meaning that my odds of lasting glory are slimming rather rapidly. And there's a good chance any namesake of mine would get the chop within decades. Half of all species are parasites, which are distinctly less appealing. Plus both species of elephant are already named, alongside the one lion species. It's not looking up for me at all.

This chapter is all about biodiversity- the totality of life's abundance- and precisely what we don't know about it. Which is a lot. Former US secretary of defence, Donald Rumsfeld, spoke of three categories of knowledge. There are known knowns, known unknowns, and unknown unknowns. Life on Earth usually falls into the latter. That means not only do we not know the answers to some of the most basic questions, we often don't know which questions to ask in the first place.

Imagine life as a toothbrush. It's a comparison I've often made to jumpstart conversations, but struggled to fully justify. Now, the average toothbrush has 2500 bristles on a space of roughly 2cm² (no, I didn't count). If on this hypothetical toothbrush, each species were to be represented by one bristle, that would mean we need a toothbrush with a head somewhere between the size of ten A4 sheets of paper and ten football fields. Does that help? I hope so, because on the whole, it seems that people don't have the faintest idea how much life is out there...

2.) Approximately how many living species are we estimated
to share the Earth with?

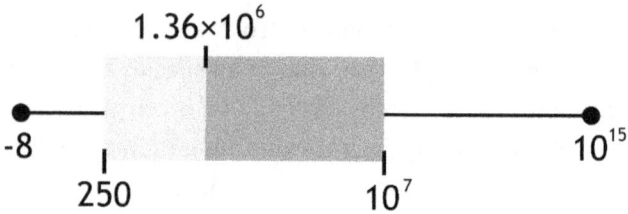

$$1.36\times10^{6}$$

$$-8 \qquad\qquad\qquad\qquad 10^{15}$$

$$250 \qquad\qquad 10^{7}$$

The two boxes each represent 25% of respondents to this question. Likewise, the two lines on either side show where the other two-quarters of people placed their guesses. The average ended up at just over 1 million. As you now know, that's a gross underestimate, so already you're learning! My favourite answer is from one person who got creative and put '-8' as the number of species overall. Go figure.

Part of the issue lies in our changing point of reference. Sailing through the Cayman Islands in 1503, Columbus described a sea full of "living stones" that made it difficult for his ships to move straight. These huge gatherings of green turtles are said to have kept his crew awake at night from the constant knocking noises of their bumping into ships. Indeed, they so impressed themselves upon Columbus that he named the area Las Tortugas after them. Nowadays, the species is a rare sighting there, hunted to near-extinction by endless fleets of vessels coming through, capturing the creatures alive to store as food. We have difficulty picturing or even believing such accounts 500 years hence. What with pollution and coastal development, global warming, and rising demand for turtle soup, we have put an end to such spectacles.

Another example to consider is that of the passenger pigeon. Flocks of these birds filled American skies as little as 200 years ago. Larger flocks could block out the sun, taking three full days at a time to pass. By 1900, none in the wild remained, with the last individual named Martha dying in Cincinnati zoo on 1st September 1914. Hunting combined with habitat loss dealt the death blow to this species, whose 3 billion individuals once comprised 1/3 of the continent's birds.

The Rocky Mountain Locust shared America's landscapes with the passenger pigeon. A single swarm of these in 1875 numbered 12.5 trillion individuals and covered an area the size of California. 30 years later, the species was gone. It had disappeared, extinct, kaput- never to be seen again. Nobody knows the reason.

This isn't ancient history or some plague of Biblical legends. This was a time when cars and phones, dishwashers, escalators and contact lenses were all coming into being. In brief, not so far removed from our modern world. And I haven't chosen American examples because they are in any way special, or to reprimand this nation in particular. It's true that the US surpasses all other countries in recorded animal extinctions, and it's true that the US scores second on having the most threatened species. But my own country Australia lies only one place behind, with a similarly patchy record. I could have told the story of the great British starling murmurations, numbering 1.5 million in a flock yet collapsed by up to 90% in a single generation. Or the Christmas Island crab, once estimated at 120 million individuals, put at risk by invasive ant colonies and losing tens of millions.

These illustrations all prove witness to a simple point: animal populations change fast. We have to keep a hand on the pulse if we're to feel these tipping points approaching. We're often incredulous towards historical evidence of animal abundance because these tales lie so far outside of our own experiences. Paying attention to such stories can still teach us important lessons about nature's vulnerability. It's custom to compare the status quo to pre-industrial levels, meaning before 1700. What we fail to recognize is that there are huge fluctuations even within such a brief timeframe and that humans wrought many changes before this chosen time. 1700 was not an idyllic world for people or for animals.

They say there are ten types of people in this world: those who understand binary, and those who don't. If you're in the former camp, chances are you didn't fail your high-school maths exams, and maybe even appreciate numbers. That's good news because it means you can read on immediately. If, however, the latter is a more accurate description of you, then I'm sorry to say you'd best skip a dozen pages and rejoin us once the science-y graphs are over. Ready?

Think back to the preface's quiz that you started this book with. How did you go? Remember that the average score was a measly 2 out of 8, with no-one getting full marks. Why did people score so badly? Perhaps they were all idiots. Perhaps I just ended up with one thousand fools who answered the exam. It's possible, except that many of those that answered were heavily interested in environmental issues too. They were well educated, followed the news, and in some cases were actively working on conservation's front lines. They self-rated themselves with an average of 7.2/10 on environmental awareness before the test, which is a fine score for a modest bunch of people.

When giving this quiz, I added in a bonus question as a sort of calibration. This was question five: "To which animal group do whales belong?" The theory was that if a large portion of respondents failed at this basic general knowledge, then I'd take the results with a pinch of salt. More than 95% gave the correct answer of 'mammals,' so I've got plenty of faith in their intelligence.

One factor that came up time and again as a cause of errors was an instinctive pessimism most people possessed. We expect the worst, sometimes embrace it as a reason to keep on fighting. However, the worst is very rarely the case. Take question one, about deforestation in the Amazon:

1.) What percentage of forest cover has been lost in the Brazilian Amazon from deforestation over the past 50 years?

A tiny fraction of people who answered got this right because we're used to hearing that the world is going to end and that everything is going downhill. We feed ourselves endless bad news and then refuse to accept it when things genuinely improve. It's like wearing reading glasses with the wrong prescription so that everything's a little blurry and blown out of proportion. The Amazon rainforest is still being lost, but far slower than was previously the case. Inversely, the question where the most negative answer was correct saw some admirable performances:

7.) Since 1970, the abundance of animals in the wild has...

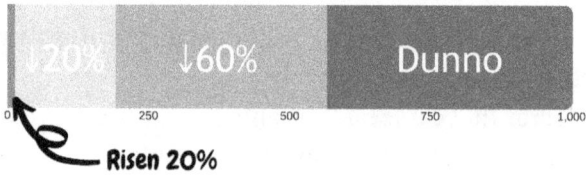

↓20% ↓60% Dunno

Risen 20%

People are also used to valuing nature, which makes any threats to it seem greater. In general, folks were more than happy to accept nature contributes a large fraction to the global GDP. Same applies with the SDGs, saying a large proportion might be impacted by the loss of biodiversity. When it came to the blame game, richer countries tended to get the hardest time of it, though in reality the average threatened species counts don't correlate with affluence. Take a look at the full results in the appendix if you're interested in the breakdown.

Climate change is well covered in news sources, but still the dramatic instinct kicks in. Media outlets have turn a profit, and stories about things turning out to be not so bad after all rarely make it in. Besides, journalists and reporters are people too, with the same tendency to focus on the negatives.

9.) If no major action is taken, how many degrees warmer will the Earth be in 2040 compared to pre-industrial levels?

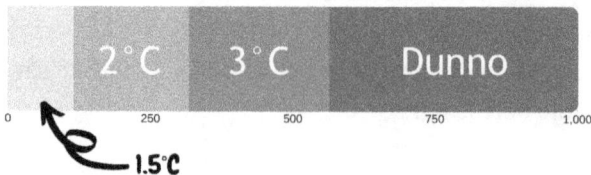

2°C 3°C Dunno

1.5°C

That in itself is not too damaging. Scientists generally agree 2 degrees is where we're headed, but an extra degree here or there seems insignificant. The danger comes when we have to prioritize these threats and set about solving them. If climate change appears like this looming killer, then won't other threats go unnoticed? So it would seem, because heightened concerns about issues like climate change and pollution led people to rank them in an order almost opposite to that which experts place them in.   The correct answer in decreasing severity should be Changes in Land/Sea Use, Exploitation of Organisms, Climate Change, Pollution, Invasive Species as you'll know from the introduction.

11.) Threats scored on severity for terrestrial and freshwater ecosystems...

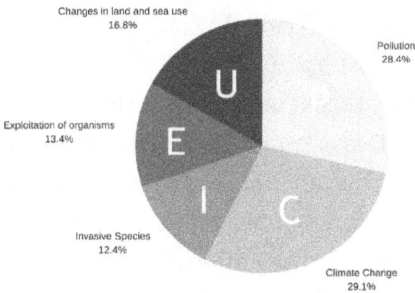

Changes in land and sea use
16.8%

Pollution
28.4%

U

Exploitation of organisms
13.4%

E

I

C

Invasive Species
12.4%

Climate Change
29.1%

Clearly we've got a lot to learn. These are the five biggest, most in-your-face challenges for our environment. You can imagine how poorly some of the smaller ones are understood.

But that's enough of graphs. Science is all well and good, but who cares? I've shown our knowledge of the natural world is as patchy as swiss cheese, and the only way I can remedy that is to answer all the really big important questions that qualify you as an eco-expert. Alongside blurry, silhouetted images of far-off Ugandan monkeys I've been sent asking for IDs, there are some gems of questions I've been asked over the years. These are my ten favourites, which really made me pause in my tracks. Have a question you can't seem to answer? Reach out, and it might just make it into the next rambling book I write!

1.) Do snakes have tails?– Jaden Lanceman

Short answer: yes. The problem of course with snakes is that their entire body is tail-like. A tail can be loosely defined as a long, skinny appendage extending from the body and opposite the head. Snakes have heads, but where does the tail start? It's not as much of a trick question as it might sound. The tail starts directly after the 'anus.' Without going into too much graphic detail, snakes– and birds– have a single rear opening that serves all purposes. It's called a cloaca. The tail is just the bone and flesh after that point. Phew!

It so happens that snakes find this question almost as hard as you did to start. Occasionally, one will get confused and attack its own tail. Cats and dogs can spend all day at this task, but as classic arcade games show, it's far too easy for snakes. The result is that they may sometimes bite or even start eating their rear end. By the time they realize this, it's sadly often too late, and stories of snakes found dead after eating themselves are scattered throughout history. "What about venomous snakes?" I hear you ask. Thankfully most of them have a degree of immunity to their venom. Just like you sometimes bite your lips, snakes may accidentally ingest small quantities of their venom from time to time, so they build up anti-venoms in their bloodstream. A full envenomation from another snake could still kill, but snake-hunting snakes like the king cobra protect themselves against this with a highly-developed venom combat system.

2.) **Are viruses alive?** - Jerry Connor

Maybe? This is a question to which we have no good answer at the moment. What makes a creature 'alive'? There's no agreed-upon definition. Instead, we typically rely on a number of characteristics to decide. Living things move, breathe, react to their surrounds, grow, reproduce, excrete waste products, and take in nutrients for use as energy. Viruses fill a few of these categories, but not many. They can't self-replicate (reproduce) and rely on host cells into which they inject their DNA. They then hijack the cell's inner workings to create lots and lots of new virus copies. It's like invading a factory that makes shop-window dummies and reprogramming it to create sculptures of yourself. Viruses don't use any energy themselves, because they can do everything they need by enslaving host cells. Before that, they're limited to floating around in the air, lodging to surfaces, or being carried in water supplies until they stumble across their next victim.

In 1946, the Nobel Prize in chemistry was won by a scientist who showed that viruses really can be reduced to an under-standing of simple chemical components. We don't consider a fire or a rusty bridge to be alive, so why might this 'borrowed life' count? Some of the largest known viruses are mimi- and megaviruses, which have recently been studied in detail. It turns out they may have some of the biological machinery required to copy themselves without assistance. Amongst the earliest lifeforms were balls of genetic material called RNA that could do just this. It seems that most viruses have since lost the ability, but perhaps these tiny giants can prove that statement wrong. And if you were to use a more fundamental definition of what it means to be living: say, evolving and propagating, then viruses tick all the boxes.

My guess is that the more we find out about these super-small creatures, the more we'll come to realize they can't continue to be held as non-living. Take for example the large chunk of your DNA that's known as 'genetic dark matter.' We call it that because nobody knows where it comes from or what it might do. Our best guess is that it's from some other tiny class of animal out there just waiting to be identified. We know already that 8% of our DNA is from viruses anyway. When it comes to describing life on this scale, we're hopelessly clueless and I don't doubt there are many amazing discoveries yet to be made.

3.) Which came first: the chicken or the egg?- Sam Goodman

Believe it or not, I once had a week-long debate on this. On the whole, I'd lean more towards the egg's side, but for fear of receiving stacks of angry reader letters, I'll state some arguments for both sides.

Starting with theology, the idea of the 'world egg' or 'cosmic egg' is a very common notion throughout world religion. The Sanskrit scriptures talk of Brahmanda, a concept which translates rougly as God-egg (Brahma being the Hindu creator God and '-anda' meaning egg). Similarly, the orphic egg of Greek mythology was said to have hatched the deity Phanes, who in turn created all the other gods. Eggs also feature similarly in Chinese, Egyptian and Finnish mythology, just to name a few.

The counterargument here comes from Christianity, which goes against the grain. On the fifth day of creation, God apparently created "every winged bird according to its kind"

(Genesis 1:21). He then gave these birds the ability to reproduce via egg-laying, which means the chicken definitely came first in this instance.

From an evolutionary standpoint, the domestic chicken was bred from the red junglefowl and possibly also grey jungle-fowl species of India and South-East Asia. This process, by definition, must have been a gradual one in which successive generations of proto-chickens laid a series of progressively more chicken-like eggs. At some point in this long progression, we reach a stage at which two proto-chicken ancestors come together to produce a single egg which a random mutation of DNA has caused to become a chicken (wherever you draw the line). Jeans can be replaced every year or two, but genes stay with you more or less the same for life. The genetic code of the chicken, therefore, had first to be seen in the chicken egg.

If you take the question at its most literal, then eggs preceded chickens by millions of years. An approximate origin of egg-laying might be pinned at the Cambrian explosion 541 million years ago, when most of the animal groups we know today emerged. These eggs would have been fish eggs, but the first eggs on land still stretch back a lengthy 312 million years. The dinosaurs that eventually give rise to chickens and birds were laying eggs 150 million years ago, and you very well may not be able to tell one of these apart from the kind you cook in omelettes today. They were laid in small groups, slightly pointed at one end, and even those of the largest dinos barely exceeded half a metre in diameter. That's enough to make a killer omelette, if farming didn't present such a challenge! Chickens themselves were domesticated between 8,000 and 10,000 years ago, so on the whole, they arrived very late to the party.

Conversely, a recent study shows a single protein is respon-sible for chickens' characteristic egg-laying plenitude. The remarkable rate at which these fowls lay eggs is a major reason why they and not flamingos were domesticated. They also don't choose caustic volcanic lakes for their nesting grounds, which may have played in their favour. Anyway, this protein is found only in the ovaries of mature females, so it could be argued on these behavioural grounds that this mark of a true chicken was first found in an adult bird and not in an egg. The debate continues...

Finally, using linguistics as a tie-breaker, the word 'egg' arrived in English before 'chicken' by most accounts. Just like a 'herd' was once used for any group of animals from birds to bats to bees, a 'chicken' originally just meant any young bird. Who knew? Also, the vast majority of animals alive today lay eggs and are not chickens, so from a majority wins viewpoint, then the answer is clear. That's how most questions should be answered, I think. Just vote on it!

4.) Are zebras black with white stripes or white with black stripes?- Arnaud Brisset

Thankfully, the answer to this is rather more straightforward. Historically, it was believed that zebras were white with black stripes. One explanation of this lies in the fact that some zebras have plain white underbellies with black stripes extending only down their sides. The natural assumption is that these black stripes have been 'painted' on the white canvas of the body below. Not scientific, you say? You might just be right.

Nowadays, we know that zebra embryos have black skin. If you shave a zebra you would find the same: the underlying skin is a darker shade. Hence zebras are black with white stripes. The way to remember this is that zebra crossings start off black from the asphalt and have the white stripes painted on. You see, the answer was staring us in the face all that time!

What are these stripes used for? Now that's a more interesting question. The answer is probably a mixture of a lot of different factors: they help cool the animal, deter horseflies, provide disruptive camouflage when running, and allow zebras to recognize others in their herd. The stripes are like fingerprints- each pattern is unique. Except that fingerprints are subtle identifiers that most people ignore. The same can't be said for stripes.

The classical explanation of a zebra's stripes is that they play tricks with a predator's eyesight. They were said to form an optical illusion, with the animals' shapes blending into each other. Sadly scientists nowadays are sceptical of this zebra magic. Most predators have poor long-distance vision, so the effect would only work up close. Needless to say, that's cutting it pretty fine on the zebra's part. More importantly, new research shows that stripes may be counter-productive for a zebra trying not to be singled out in a chase. It seems the pattern might actually help the attacker to follow a target's movements. Oops.

The theory of motion dazzle- where stripes break up the outline of a moving object- was widely used in the 2nd World War. Some battleships were painted with zebra stripes, and canons too, without the supposed benefits of the high-contrast pattern being tested. I'd like to see pink polka dots on missiles by the time World War 3 comes around.

5.) If a tree falls in a forest where no-one can hear it, does it make a sound? - Adam Kagan

I hate this question, but I'll try to be polite. The answer is yes (obviously). The core of the puzzle is that sound by definition relies on its being perceived by a listener. In brief: a sound is not a sound until it's heard. What the question blatantly ignores is that there are creatures with good hearing in every forest on the globe. The trees have ears, so to speak. To prove it let me illustrate with Chernobyl's Red Forest.

When the nuclear reactor blew at Chernobyl in 1986, the surrounding area was flooded with lethal radiation. So you might be forgiven for thinking it became a total wasteland. Months afterwards, many of the animals it had formerly sheltered were nowhere to be found. Note 'months,' because this event happened over 30 years ago. Nowadays, the situation's a little different. In the famous Red Forest of Chernobyl that boasts some of the highest radiation levels, dormice are thriving. That means there are hundreds or thousands of tiny pairs of ears scattered throughout the woods. Birds have returned, and a bunch of other animals that call the place home. Without decomposers breaking down the leaf litter, the autumn drifts accumulate in piles many feet deep. Presumably then, any tree that did fall there would make quite a racket.

It's clear that every forest has little critters strewn throughout that serve as its ears. If any forest were to be mute, Chernobyl would be the place without sound. It's not. In fact, Chernobyl is a remarkable case of rewilding, where animals have reclaimed the land after humans moved out. There are wolves, wild boar and horses, hares, foxes, moose, brown bears, lynx, pine martens, otters, beavers... all manner of creatures.

Smaller animals are still struggling somewhat with radiation levels, but a remarkable ecosystem has sprung up in their absence. With a 4000km² exclusion zone surrounding the site, there's an unprecedented opportunity for animals to live outside the reach of human interference. Areas are being sectioned off and recognized as national parks: 'radioecological' zones. That's one word I never thought I'd hear.

6.) Is there really no such thing as a fish?- Aurelie Amiault

This "fact," now the title of a leading podcast, is often attributed to the great biologist Stephen Jay Gould. Gould is known mostly for refining the theory of evolution, but also took a major interest in fishes. *The Oxford Encyclopedia of Underwater Life* states:

"Incredible as it may sound, there is no such thing as a "fish." The concept is merely a convenient umbrella term to describe an aquatic vertebrate that is not a mammal, a turtle, or anything else. There are five quite separate groups (classes) of fishes now alive – plus three extinct ones – not at all closely related to one another. Lumping these together under the term "fishes" is like lumping all flying vertebrates – namely, bats (mammals), birds, and even the flying lizard – under the single heading "birds," simply because they all fly. The relationship between a lamprey and a shark is no closer than that between a salamander and a camel."

That pretty much sums it up. You could draw parallels to vegetables as another defunct grouping. Tomatoes grow from flowers and contain seeds, which are the two tick-boxes that define fruits biologically. Yet the US Supreme Court has ruled that tomatoes are vegetables. Apart from the obvious question: "Don't they have anything better to do?" this raises aan interesting point. It shows that whilst groupings can be fairly arbitrary, they're still important. For tomatoes, it was a matter of taxation. For fish, it's a matter of life or death.

Mostly, we have cuisine to blame. Throughout the 16th century, everything from seals to whales to crocodiles and hippos was called 'fish.' Cuttlefish, starfish, crayfish, jellyfish and shellfish still are, despite looking nothing like real fish. Many of these are listed alongside and in-between cod, salmon, herring and bass in seafood menus. To add to the confusion, when you order sardines, you might be consuming any one of twenty small fish species which are similar in shape and texture. Between one-fifth and one-third of all seafood sold is mislabelled, so chances are you don't know much of what you're putting in your mouth.

Then there's the church- the Catholic church, to be precise. During Lent, followers of this religion are forbidden from eating meat on Fridays. Except for some reason, fish are exempted. So by pure coincidence it happens that Catholics have a *very* loose understanding of what is meant by a fish...

The capybara is a giant South American rodent (the world's largest) and is considered a "fish" for Lenten purposes. When the Spanish settlers arrived on the continent, they wrote to Rome describing the animal. The answer was a papal declaration saying that as it swam, it was definitely a fish. In 2010, the Archbishop of New Orleans declared that alligators were considered part of the fish family too, and puffins have long held their place in this list. Beavers were added to the menu by the 17th century Bishop of Quebec, whilst in Nicaragua, lenten meals include iguana, armadillo, and shellfish soup. It'd be shellfish to leave them out now, wouldn't it?

Superfishally, the characteristics that define a fish can be listed. They live in water, have a backbone, and possess several sets of fins. They are also cold-blooded... except a few of them which aren't. And you have to discount other marine creatures like dolphins, manatees, porpoises, whales, turtles etc. etc. etc. The whole system falls apart- there's just no easy way around it. Grouping them by a single common ancestor would result in us being fishes, alongside all other land vertebrates. Best mullet over some more!

7.) Do butterflies remember being caterpillars?- Jonkelthan Famofan

I'll admit it: this is a personal favourite. You'd think it impossible that butterflies could retain any of their memories after liquidizing themselves in metamorphosis. Yet they do. When the caterpillar enters the chrysalis, its body turns into soup and slowly rearranges itself into the new configuration required to begin life afresh as a butterfly. That in itself is a mind-blowing feat, because it means that one animal can take advantage of a dual life. In its youth, it feeds on plant leaves, and as an adult sips on nectar and fruit juice. It starts off sluglike and slow moving, and ends up an adept flier. It's a bit like turning from a fish into a bird, only fast-tracking hundreds of millions of years of evolution in the process. Wow.

Recently, scientists trained caterpillars to avoid the smell of nail polish remover.They claimed it was for research. When they (the caterpillars) had undergone their metamorphosis, they were able to remember perfectly well this simple principle. The adult butterflies knew that nail polish smell meant some- thing bad was coming, and so steered clear. Only if they were taught very early on in their lifespan did they forget, just as you probably don't remember your first Christmas or very early childhood. It's possible that this helps adult butterflies and moths to lay their eggs on the right plants for their offspring. Remembering what they ate as a caterpillar allows them to put their babies in the larder, as it were. If we work out how this memory retention is managed, it would mean a major breakthrough in human medicine, with applications in treatment for victims of brain damage and disorders.

8.) Why do monkeys like bananas? - Andrew Tang

Mostly for the same reasons you do. They taste good because they've been bred to taste good and are full of delicious sugars. Despite what Hollywood would have you think, monkeys do not encounter bananas in their natural habitat. The banana plant (technically a herb) comes from Papua New Guinea, where monkeys are not found. Only recently have humans released invasive monkeys on the island, which doesn't really count. And wild bananas are a far shot from the bendy yellow fruit we know, on the verge of being inedible. That aside, bananas are a big favourite with zoo animals- so much so that keepers must take care they don't overdose on the fruit. Melbourne Zoo in Australia famously banned bananas for its monkeys after discovering it led to obesity and tooth decay for the primates. Other zoos have followed suit, though the monkeys weren't consulted on the matter.

Great apes like chimpanzees follow a similarly strict regime in institutions. A week's chow might include beans, beetroot, broccoli, bok choi, cabbage, capsicum, carrots, cauliflower, celery, corn, coconut, cucumber, endive, eggplant, eggs, leeks, lettuce, spinach and shallots, tomato and turnip. A true vegetarian buffet, though lacking on the fruit front. Dare I say, McDonald's certainly could take a few hints for its menu!

It's not clear where the whole monkey-banana myth originated, but what we do know is that grapes take first place when it comes to preferences. They're even more sugary, so it's hardly a surprise. Incidentally, if you're wondering how monkeys peel bananas, the truth of the matter is they don't. Usually, they'll just take a whopping great bite out of the middle, eating the whole thing skin and all. Had humans followed their example, much of comedy's glorious history would likely have been lost.

9.) Is the Loch Ness monster real?- Zac Fernandez

Despite some valiant efforts to affirm this, science remains sceptical. The modern legend kicked off in 1933, when a now-famous photo of a large grey creature was captured rising from the lake. Nessie-deniers say it shows an elephant from a circus that was passing through at the time, but annual sightings continue almost to this day. Circus elephants not so much. 2013 was marked out as the first year where no half-decent photographic evidence arose, bringing up major concerns for Nessie fans. Rumours spread that their favourite friendly cryptid might have given up. The fears were unfounded, of course, and Nessie has since re-emerged.

An annual prize of £1000 has been set up and is awarded for the most promising sighting of the Loch Ness monster each year. It's sponsored by the betting agency William Hill, which offers gamblers odds of 500:1 for the Natural History Museum confirming Nessie's existence within one year of the bet being placed. Some would say that's a niche betting market, but perhaps some Nessie fans will put forward the money?

In 2005, 100 international athletes were each insured for £1m against loch ness monster bites during the swimming section of Scotland's largest triathlon. As justification, the firm responsible said "The competitors will all be very psyched up and very driven, so there's going to be a lot of noise and a lot of splashing going on. Just the sort of thing that might annoy a prehistoric lake monster. Or even worse, give it an appetite." Nessie supporters were understandably miffed: "Everyone knows she is friendly; she has been present in the loch for centuries and never hurt a soul in all that time," said one local. "If she wanted to get involved in the race, though, that would be another matter. Through her knowledge of the loch, she'd beat everyone. She might even continue on land and finish the competition." Fancy that. Fortunately, the event passed without Nessie making an appearance or attacking any of the athletes.

In recent times, the 20th-century polymath Robert Rines became embroiled in monster-mania. A Harvard lecturer, he somehow found time to continue the hunt for Nessie alongside an impressive list of feats. These included playing a violin duet with Einstein, filing 800 patents (including one for hinged chopsticks), and helping to find the titanic wreck in his spare time. He brought similar pizzazz in his search for cryptids.

He hired a perfumer to concoct an odour that might be attractive to Nessie, and later trained dolphins to carry cameras on their heads on a search mission for the creature. This was inspired at large by a trip in 1972 where he claimed to have sighted the beast. However, after many decades of trying, he had only a handful of poor-quality photos as evidence of his labour. The closest anyone has come to contact is when a 9m model of the monster was found at the bottom of the lake, having been discarded there from a 1969 movie set. Such a shame.

If there is a silver lining to Nessie mania, it's that the loch ness monster has indeed contributed to saving living creatures. A New Zealand scientist named Neil Gemmell is pioneering a technique called eDNA. This involves taking water samples and analysing DNA fragments they contain to discover aquatic lifeforms living in an area. The hope is that this will provide much more accurate data for monitoring fish stocks, but funnily enough people aren't interested in that. For a publicity stunt, he went off to Loch Ness and sampled eDNA for the monster. His quest aired on a 2-hr TV doco, again without a single Nessie appearance. The findings of his testing suggest that Nessie is most likely a giant eel. Though no-one knows what monster DNA looks like!

10.) What do you call a purple armadillo?- My Future Self

A bad paint job.

3

The Commonwealth of Birds

P eople are rarely surprised when I say I'm British-born. Perhaps it's the tottering piles of tea mugs on my desk, or my partiality towards castles. Whatever the case, something always seems to give it away. Sure I've picked up a few Australian-isms in recent years, but having avo and toast in the arvo just doesn't cut it, and vegemite's disgusting. There is something about being British that stays with you for life, and for me that means bird-watching.

Looking from the outside in at these sodden isles with their life-sucking winters, you'd be forgiven for thinking that sitting around ovens watching roasts cook would be a more suitable past-time. It is. And yet, by some fluke of nature, the British have taken on a most extraordinary trait in their love of birds. Survival of the fittest surely cannot come into play, as even the shortest expedition into the countryside a few weeks either side of high summer can prove fatal. Sexual selection? Possibly- for what is more attractive than a man dressed as a bush peering out from between two hedgerows?

Sadly, science (as usual) has to intervene. In 2016, an

American professor conducted a survey of some 1300 people, described as the first true investigation into what we see as 'creepy.' There were the usual suspects: clowns topped the list, followed by taxidermists, sex shop owners and funeral directors. But there was also an entry that you wouldn't expect on the leaderboard. That's right- birdwatchers.

On closer investigation, there might be some clues as to why. Take the traits people listed as being creepy. These included wearing dirty or weird clothes, displaying too much or too little emotion, talking too much about a topic, smiling peculiarly... it's a long list. If you're a bird-watcher, you're probably nodding knowingly at this point. If not, you'll just have to take my word for it. I promise you I fit all of those criteria and more.

Self-pity aside, where is all this leading to? I believe there is a third much more critical factor that explains the habits of the bizarre British public. What on Earth could it be? Now, as I'm writing this, Birdlife International's database lists a total of 267 bird species for the UK. That's pretty rubbish in relative terms, ranking it 149th of all countries for avian biodiversity. By comparison, Australia has a highly respectable 724, placing it 33rd. Numbers vary of course, but having almost triple the species count has got to mean something. Add to that the fact that Australia's birds are objectively better- more interesting and beautiful- and the plot thickens.

My explanation of British bird-mania lies in the most funda-mental of conservation measures: people. The Royal Society for the Protection of Birds (RSPB) is Britain's major birdwatch-ing collective. It's also the single largest conservation charity in Europe. That's no mean feat, given that this continent lies at the epicentre of environmentalism globally. With over a million members in the UK, the RSPB possesses a role of unique importance in promoting this hobby.

The equivalent Down Under is Birdlife Australia, with the rather-less-grand total of 12,000 members. I've worked with most of this organization's staff, and can tell you they are some of the most passionate people you could hope to meet. The difference between this and the RSPB is not about the people behind the wheel, so why this huge discrepancy? The RSPB was founded in 1889 by a pioneering female activist whose name was Emily Williamson, and who at the time was campaigning against the use of bird feathers in fashionable ladies' hats. That makes it comparatively ancient as an organization, but Birdlife Australia has been running under various names since 1901, a mere decade later. No great difference there. We have to look back further.

Naturalists are like geriatric rabbits: they pop up just about everywhere, but they're dropping dead almost as fast as they can grow their numbers. In affluent areas where healthcare is good and food plentiful, they tend to survive longer and their traits are continued by successively larger generations. Hence the stereotypes of doddery Victorian gentlemen peering at the world around them through a monocle. These were the early naturalists and explorers, who paved the way for many of the early scientific breakthroughs.

Nowadays, birdwatching is the second-fastest-growing hobby in the United States, and similar trends can be observed in both Britain and (to a lesser extent) Australia. Evidently, a greater initial number of followers (or rabbits) means a faster subsequent rate of growth. And it would seem that this has given Britain the upper hand in popularizing birding, but then again, the chicken has to hatch from something. Why are

British roots for the hobby so well-grounded?

Feeding and feasting. That's the cause. Since way back, chilly Poms have left out chili-laced seeds to feed their feathered friends whilst warding away any scavenging squirrels. It's a habit that, once started, is very hard to stop. Birds aren't stupid, and routine rapidly becomes obligation as an unspoken contract is made on both parts. The little tits and robins turn up every morning, and in due course, breakfast appears in the bird-feeder. If dogs hadn't claimed the title of man's best friend, the plucky persona of the robin would have placed it well in the running. Such is the bond that bird-feeding forms.

In Australia, bird-feeders are shunned by society with naturalists warning of the awful consequences of the practice. Some deep-seated belief of obscure origin has seated itself in the minds of the public that means feeding these creatures has become anathema to most. Science itself is ambivalent. Nothing clearly shows large positive or negative effects of the practice. Most likely it's a sentiment carried over from the earliest white colonists on the continent, who saw local birds as a loud nuisance. And whilst lorikeets' beaks from long experience I can tell you are a precedent to thumbscrews, you're told not to feed birds for their own sake, not to conserve all your fingers.

Feeding birds allows your everyday Joe or Joeline to appreciate and connect with local wildlife. The people are the real winners and not the birds. Learning the names of the regulars that appear in your bird-bath is a crucial first step towards the world of birdwatching. But there's one more element we've missed. What about feasting?

This is where things get both incredibly complicated and incredibly interesting, because feasting goes waaay back. In feudal England, kings protected great swathes of land for their personal hunting use and to provide meat for their table. Lords did likewise, and later gentleman collectors shot and stuffed specimens for study, forming the basis of natural history disciplines. In Australia, convicts arrived in 1788 and set about clearing the land for their own purposes (mostly farming). In doing so, they displaced the Indigenous peoples who had managed it for eons.

Hunting and collecting birds are still damaging practices just as converting forests to farmland is. The difference is that in Britain, the manner in which these activities were performed helped to set useful precedents. In Australia, not so much. Whilst UK hunting grounds became nature reserves, Aussie farmland became houses and roads. Studying birds, even though it required shooting them, allowed the public to learn more about the creatures, and grow a certain fascination. Developing recipes for parrot pie as was common in Australia did not have this same consequence. That's all there is to it.

Enough about birds. This chapter is about the challenges facing conservation as a whole and hence why we need to change our viewpoint so drastically. My opening anecdote is a mere illustration of some of the discrepancies contained within the field. What does the broader picture look like? What of this looming future we try not to confront?

As I write this, I'm looking at Google Trends and seeing the term 'Tiger' trending in the US alongside Boris Johnson, Kourtney Kardashian and Pope Francis as a popular search. That's strange- an animal that's trending? Impossible. I click through and discover the word Netflix pasted all over the popular sites. It seems there is a particular crime series called Tiger King hitting the screens. Okay: that's not exactly what we're looking for. So I try 'lion' and, funnily enough, there's a huge peak in worldwide searches mid-2019, when Disney's The Lion King was released. The same can be said for 'clown fish' with the 2003 Finding Nemo release, and pretty much any other example you might care to mention.

This is to be expected: we live in a digital age, and productions whether they're Attenborough documentaries or animated films give animals an enormous boost in publicity. The recent Lion King film was revolutionary in going one step further, redirecting a slice of revenues right back into conservation schemes. Good on Disney for doubling the favour.

The fun with Google Trends goes further in comparing different species. Accordingly, elephants can be seen to be ten times more popular than termites, and dogs beat cats by a factor of two. In the abstract, the term 'biodiversity' does okay all-round, but plummets annually in July and December. 'Conservation' follows a similar trend. Perhaps family time and festivities intervene, but this phenomenon is quite unique to the terms. It must be a matter of priorities- you don't look up rhino poaching to have a good time, and conservation has always been depressing.

The mega-trend that emerges throughout all this is that certain animal species have a habit of stealing the spotlight. Rhinos, tigers, elephants, pandas and lions all count on this list, and have firmly seated themselves in our hearts and minds. They appear frequently in popular culture, and conservation recognizes this potency. They're termed 'flagship species,' because they act sort of like ambassadors to the rest of life on Earth. They attract more donations to charity, spark public debate, and look good as mascots for sporting teams or on the side of cereal packets. Koalas perform a similar role here in

Australia: they're adorable and furry and look rather like teddy bears, even if they refuse to pose for photos. The difference between koalas and other flagship species is that protecting them helps plenty of other animals. Neat, right?

Whilst China's ecosystems can do without pandas (tragedy though their absence would be), failing to protect koalas in Australia jeopardizes the environment. They are an example of what is creatively termed an 'umbrella species,' not because they're especially water-resistant, but because protecting them means protecting lots of other animals as well. The tricky thing is that most umbrella species aren't adorable or widely loved. So we have a dilemma: either soak up what little funding is available by focusing efforts on big charismatic keystone species, or change tack and work on those umbrella species through which conservation can make the greatest impact.

There's not really a choice. Taken in one lump, conservation has an annual budget of about 25 billion USD to work with globally. That's about the cost of hosting the Olympic Games, and less than the Croatian government's annual budget. If you do the maths, it's just over 1 US cent per billion animals on Earth. I'd argue that's stretching things a little thin. Charities right now need to maximize their funds to create any impact whatsoever, and governments fighting to hold office need the public's votes just as much. You might get brownie points for protecting critically endangered vampire spiders, but I assure you that doesn't equate to wealth.

Quite the paradox it would seem. Here's another one to muddy the waters: our political system is founded upon debate, but does talking really equate to better outcomes for animals? When the UN Climate Action Summit took place in September of 2019, web searches for climate change skyrocketed. Sadly other events aren't quite so media-worthy and don't create so much as ripples in the press. In policy terms, goals tend to look good on paper without often being met. And when governments intervene on environmental matters, the issues themselves become politicized in ways that don't help anyone.

Ironically, countries with more non-profit researchers have substantially more threatened species, whilst those with a higher proportion of journalists tend towards having fewer species threatened. There's some very old rhetoric that we shouldn't spend billions on space exploration whilst destroying the one habitable planet we know of. That makes sense, but it turns out that countries with a strong aerospace industry have on average much stronger environmental policy. Causation might be a stretch.

These sweeping claims come from a mega data set I amassed recently, looking at 87 environmental and developmental factors for the world's 192 countries. It was an exercise driven by curiosity with only a trace of scientific rigour. But of all the thousands of insights my analyses revealed, one stood out to me in particular. Countries with more of their area protected for animals were losing species just as fast as those with less. Strange, right? Isn't habitat protection and more wild spaces what tree-huggers (myself included) have so long campaigned for? More forests, more national parks, less urban sprawl. And yet it seemed that places with stacks of these pristine areas weren't protecting anything after all. That's not true (thank God), but what it does show is that not all protected areas were created equal.

I've always had mixed feelings about Antarctica. True, it's meant to be the last untouched wilderness, but I've never quite believed those stories. It's still got some 200 threatened species, which whilst not a huge count comes from a relatively small pool of animal residents. In 2019, stories hit the media about a major penguin colony collapse: 25,000 emperor penguins had dispersed after an unsuccessful breeding season. That sort of thing shouldn't happen in the natural order of things. The challenge stems from the fact that many countries have claimed a chunk of the continent for themselves. Almost all of Antarctica is protected in some form or other, but because the nations responsible form such a complex patchwork, conservation is both everyone's and no-one's responsibility. Protected areas are all well and good, but they don't protect themselves.

Then there's Australia, my adopted country. It has the third most threatened species of any country (7,618) yet boasts a whopping 30% of its lands and oceans protected. At last count, global averages were about 15% of the Earth's land surface and 7.5% of the oceans protected. Australia is the 12th best country on this score. The thing is, Australia is mostly desert: hot, barren, and as close as it gets to lifeless. Look on a map and see for yourself- it's hardly prime real estate. Lots of the really big chunks of protected land are found here. Some would call that cheating. But even in the nice lush national parks, authorities don't have the means to enforce Australia's patchy environmental laws. Land, after all, is just land until someone's going to the trouble to look after it.

Not so long ago in a galaxy not so far away, the most brilliant biologist of our time- a certain E.O Wilson- put forward a vision for a 'half earth project.' This was not the result of a misfire from the death star, but rather a nifty idea for divvying up Earth's limited space. Having performed umpteen calculations and made lots of complex models, he showed that by carefully selecting and properly preserving half of the Earth's lands and oceans we could save over 85% of species from extinction. If you're not impressed, you should be. That's millions or billions of species saved, as opposed to the 50% extinction rate we're currently headed towards. The key to an idea like this is picking the right spots to protect: choosing areas that represent the greatest possible range of species. And not forgetting about the protecting part, either. Learn from Australia.

Recently, National Geographic surveyed 12,000 people on the issue. The majority of them said that more than 50% of the Earth should be protected. So why isn't it? Unfortunately, people grossly overestimate our current progress in this respect, believing on average 35% of the Earth to be protected. That's a far shot from the true values I quoted above, and it means people are dangerously ignorant when it comes to policy. If we had protected 35% by now, we could perhaps afford to be more complacent, and the leap to 50% mightn't be so great. The reality is that we've got ten years to make it to 30% and need 50% protection by 2050 in order to meet with the SDGs and Paris climate goals. It is not going to be easy.

To give another brilliant analogy, you can think of it like a piñata at a kid's birthday party. When we play with policy, we're changing the shape and structure of the cardboard shell, altering its protection of the nature inside it. The larger the piñata (i.e the more land we try to protect), the faster it'll collapse when we start beating the crap out of it. Bigger piñatas are easier to hit and knock chunks off, so taking active measures to ensure these protected areas stay protected is important. If legal protection or enforcement is flimsy, you can imagine the piñata as being made from soggy cardboard- it's going to be shredded an awful lot quicker. Finally, if we're hitting a line of bunting instead of a big cardboard box, then the shock of each blow can be better absorbed across flags on the string. This is the benefit of linking up protected areas in habitat corridors, making it much harder to disembowel the whole thing. Really it's a wonderful comparison (he says modestly).

So we need more land area protected, and the right land too, in the right areas with the right measures in place to safeguard the animals within. Is that all? No, it isn't. Otherwise all anyone would be doing is buying up land like there's no tomorrow and trying not to harm it. No- one of the largest issues is not so much lack of land to work with but lack of people power in the form of public awareness and involvement.

Remember my handy Google trends data earlier? When I look up terms like 'environment,' 'conservation' and 'biodiversity' they all show graphs of identical shape. 'Climate change' is the opposite. Whilst the trio above are all smooth curves, changing little over time, climate change searches strongly resemble the contoured graphs of CO_2 emissions. The search numbers produce jagged peaks and troughs, but show a trend in the issue rising steadily in the public psyche. In an ideal world, these terms would complement each other slightly: mass media from the latter helping to raise awareness about broader environmental concerns. Only it isn't. What's going on?

Part of the problem lies in the 'green' spectrum. I'm not the shouting protester type, though I've certainly played a part in such events. Likewise, I'm a little young to take on David Attenborough's role, berating inaction from behind a creased face at press conferences, adding his sombre tones in commentary to images of forests being destroyed. Both of these cases show clear roles to be played, but they fit at two ends of a wide scale. The 'green spectrum.' Most people aren't especially vocal nor are they especially authoritative, so cluster in the middle of the range. However, with changing conditions, activists are being pushed to fresh extremes and even the most conservative public figures are being forced to take a stance.

Climate change affects humans in immediate and obvious ways, hence those pushing the issue have traction. As for invasive species, hunting, fishing, farming practices, pollution, and all the rest, they're represented by the mass of concerned citizens in the middle of the scale. These people have the numbers to enact change, but lack the direction and means of expression by which to do so.

Let's put people and politics aside for a minute and focus on some of the more fundamental problems we face in protecting our planet. Firstly, our understanding of nature has more holes than a pub dartboard. They say that curiosity killed the cat, but scientific endeavour had had several centuries to wipe out felines, and, by and large, it has failed to do so. Despite the occasional weapon of mass destruction, and some not-so-pleasant ends to the scientists themselves (radiation exposure, chemical poisoning, and being blasted to bits) the world is better off because of our curiosity. So why do we know so little about nature?

Because we're only human, in the most literal sense. We're limited to what we can observe, and that means for the most part if we can't see it, then to us, it doesn't exist. If a tree falls in a forest and no one is around to hear it, does it make a sound? Yes it does. We simply don't know about it. In the same way, if we can't find a species, we've got our work cut out for us protecting it. Introducing the Lazarus taxon. These are animals we thought were extinct and then were found again: in short, some of the most embarrassing stories known to science, where animals were literally rediscovered from right under our nose.

The so-called 'false killer whale' was first described in 1843 on the basis of a 126,000-year-old skull. Dead, that is. The scientist examining the specimen- none other than the inimitable Richard Owen from chapter 1- therefore made the reasonable assumption that the species was long gone. Twenty years later, one recently deceased washed up on the shores of Denmark. Øv! Din satans nar!

Then there's the coelacanth: not one of the most beautiful fishes, though probably one of the most stubborn. Known from fossil evidence dating back 66 million years, it again fell into the basket of almost-certainly-extinct until a fisherman on a remote island pulled one up from the depths. Now, unless it's got a museum plaque, your average fisherman has precious little chance of recognizing a fossil fish. So it was extremely fortuitous that a passing biologist took a liking to the specimen, stored it in the freezer of a local mortuary, and finally took it back to a museum for further study and preservation. That was in 1938, and since then only a handful of others have been found.

A coelacanth is no small fish, coming in at a brutish 2 metres in length and much bulkier than your average piece of hand luggage. When a cranefly species was rediscovered by landing on a scientist's hand, that was understandable. But when a species is larger than a suitcase and hiding in plain sight, you really begin to doubt our human abilities. Take the Fernandina Tortoise as an example. This is one of the Galapagos' giant tortoise species, all of them at least ten times over the weight allowance for check-in baggage. Last seen in 1906, you would have thought it had nowhere to hide on the slopes of a barren volcano. You would be wrong. In February of 2019, a film crew covering supposedly-extinct species stumbled upon a lone female making an unconvincing attempt to disguise herself as a bush. Since the discovery, $100,000 USD has been placed behind attempts to see the species conserved.

Once a lost animal re-emerges, it's all fame and fortune, with everyone fighting tooth and nail to see it stay alive. Remember though, we've still only found one of these tortoises, and a centenarian at that. If we're to get a new generation of tortoiselings... well, it takes two to tango. One of the most romantic stories of 2019 covered a frog called Romeo, thought to be the last of his kind. Romeo was kept in a cage in Bolivia until finally a 'Juliet' was found for him- five more of the species brought into care. The team responsible has high hopes that Romeo can hit it off with one of the females, although first approaches have been slow. In cases like this, a species can usually make a comeback. The question is: what happens to those species that don't get noticed, slipping silently into extinction?

Batrachochytrium dendrobatidis, known to the less scientifically inclined as the deadly chytrid fungus, has decimated amphibian populations across the globe. The disease was spread through human travel, evolving over time into a new more virulent strain. The resultant pandemic has caused millions of deaths. Sound familiar? It's been called the "silent killer," preying on frogs and salamanders by "eating their skins alive." Trust me- that's not a pleasant way to go. Most threats of this kind are relatively small. They're unable to survive for long outside of a host, typically slow to spread, and able to infect only a handful of species with limited damage caused.

Bd chytrid (as its friends call it) is special. At least 700 species are prone to infection, of which 500-odd are experiencing marked population declines. 90 of these have gone extinct in the wild, and a further 124 have fallen by over 90% in number. The chytrid spores can survive for months, possibly years outside of frogs, and even swim through water to find new victims. Some species aren't adversely affected, so comfortably carry it and spread it even further, faster. Those that aren't so fortunate rapidly succumb to cardiac arrest.

The truly worrying thing about chytrid is how long it took us to clock onto this deadly pathogen. It was first described in the 1980s, when it was already spreading around the globe. Outbreaks have occurred on every continent except Asia, where the animals naturally live with other chytrid fungi strains. South America is especially badly affected, whereas other regions are only now witnessing its advance. We honestly can't say when the disease emerged, because no-one was watching out for it. Large die-offs in the 1950s and -60s very likely have chytrid to blame, but researchers suspect the deadly fungus emerged in the early 1900s.

Right now, it's firmly rooted in frog habitat globally. Fungicides could remove it, but not on such large scales. So we're faced with the uncomfortable situation of having to accept it, limit its spread from here on in, and try where possible to reduce other pressures on amphibian populations to maximize their chance of recovery. All known strains of chytrid have been found in pet-shop animals, which points to a very likely cause of its spread. We need tighter biosecurity protocols, but we needed them a century ago.

This helplessness is enough to make even the most light-hearted conservationist feel distinctly uncomfortable. Thankfully, monitoring today is much better and chances are we'd pick up deadly threats like this fairly quickly for containment. We hope. In cases like this, where the threat crosses national borders, we encounter whole new suites of challenges. Whose responsibility is it to prevent a deadly fungal disease? The country in which it starts, perhaps. And say it has crossed a border- then another country's negligence is causing problems elsewhere, with the potential for global harm.

All of these countries are partly to blame because their border security measures failed to prevent the disease from passing into their ecosystems. The tendency, however, is simply to play the blame game, and not take any responsibility. "My country didn't start it, so it's your job to fix it!" Except shouted incoherently across Parliament with poor grammar and a litany of further complaints in the manner preferred by most politicians. Intergovernmental organizations can do great things, but they still rely on their member countries to agree on actions, and to act on decisions made for proper enforcement.

All of which brings me to one of the most damaging popular misconceptions of our world today: Africa Syndrome. The fourth question of my little test at the beginning was about poaching in South Africa. I won't tease you with trying to answer it again, so here are the results:

4.) How has the number of recorded poaching incidents of rhinos in South Africa changed over the past 5 years?

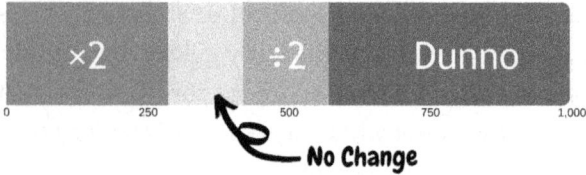

The right answer is the most hopeful one, which is opposite to what most people guessed. When I tell responders this, some of them complain that they were trying to second-guess the test. They claim that they picked the most alarming answer because that's what I'd want to be telling them. Firstly, I am very happy that poaching numbers have been dropping as steeply as they have, and I'd <u>much</u> rather inform that they've done so than drive home some false pessimistic message about the world on the verge of collapse. Secondly, I think there's more to it than that.

Part of it is the perspective we've been given by dramatic media sources and desperate charities that still very much need our support. We're led to believe that the situation is getting much worse, which is true about most environmental issues across most countries. Most, not all. Because of this, we fail to recognize when some things are actually getting better. The first question about the Amazon's deforestation has a similar principle at play. Except that there, people preferred the halfway option instead of going for the all-out world's-going-to-end response. So why are the poaching question results so especially rubbish? Again, I blame it on Africa Syndrome.

If you hadn't heard of this effect before, please be reassured that very few have because I made it up. Well... made it up in that I coined the term. The effect is still very much there. Let me explain. Africa has some of the greatest challenges of all continents when it comes to conservation. When you picture it in your head, you probably see a mixture of lions, rhinos, and starving children- tigers too if you're particularly uneducated. It's true that a lot of the pressure on Africa's ecosystems does arise from interactions with people. This means we have a much easier job understanding the threats. These sorts of stories register prominently in our consciousness because the animals are large and charismatic, and the people are suffering or causing suffering in turn. News isn't news without a human element, after all. Think about it: would you rather read about an invasive ant colony on a remote Pacific island or a Maasai tribe that's fighting against lions?

The result is that we perceive Africa to be teeming with life, and losing it just as fast. Our view is distorted just like when you're standing in an elevator with mirrors either side and seeing your reflection an infinite number of times receding into the distance. More than any other region, Africa is seen through this exaggerated lens. The continent is not so greatly disadvantaged in developmental terms (it's catching up VERY quickly) and the environmental management it practices is something to be admired. They say that 'when the going gets tough, the tough get going,' and boy are African conservationists tough. Let me illustrate.

In 1977, civil war erupted in Mozambique- a bloody conflict lasting 15 years until the early 1990s. The country's most famous national park, Gorongosa, saw its wildlife decimated. Ivory was harvested from elephants' tusks to pay for new armaments, larger animals were massacred for meat to feed hungry soldiers, and the park itself was used as a battleground.

Many species suffered declines of 90-99%, and post-war population counts were alarming: 15 buffalo, 6 lions, 100 hippos, and a handful of wildebeest. A decade later, the Gorongosa Restoration Project started. The result is a world-class reserve, with increases up to 1000% in animal populations since the war. There are over 1000 buffalo, more than 600 wildebeest, and 550 hippos. That's a remarkable turnaround few countries could match.

The benefit of direct human threats is that they are relatively straightforward to combat. Human-lion conflict occurs over much of Southern Africa, where lion habitats have been shattered into small fragments and prey is scarce. Here, when a lioness takes a cow, farmers lose a large portion of their livelihood. Many governments offer compensation for these losses, but there's tremendous variation between the amount compensated and the efficiency of the service. More often, farmers take justice into their own hands. Tribes like the Maasai in Kenya have a long heritage of hunting lions, and they're brutally effective with only clubs and spears. Yet it doesn't take a genius to see the solution.

Firstly, the compensation is improved. In some parts of Kenya, an independent predator compensation scheme has been set up. If a livestock kill is reported, an officer meets with the community. Provided that no predator was harmed in retaliation, a settlement is paid, however, if any one of the surrounding communities in the scheme harms the animal, all payment is withheld. There is a very strong social and financial incentive to coexist with lions. Other programmes select and employ 'Lion Guardians' from the local community, giving traditional hunters a stable livelihood whilst appropriating their skills. The guardians help to reinforce protective stockades, track lions to warn nearby herders, and find any lost cattle that may have wandered off. These people are fit, skilled trackers with a valued position in their societies and intimate local knowledge, so it makes perfect sense to have them as part of the solution. They're simply the best people for the job.

One final solution I will mention here, that has only recently been proposed. A few years back, a trial was held for a new and possibly revolutionary stratagem. The test subject was a group of sixty-odd cattle, half of which had the unusual addition of a set of eyes painted on their rump. The theory was that lions would not attack a cow if they thought it was watching them. As the weeks passed, three of the non-painted cows were picked off by lions, whilst none of the painted cows died. It appears that this crazy idea might just be working, but it will take a lot more time and a lot more cows until we know for sure.

Hundreds of thousands of dollars annually are lost in live-stock predation from lions, but an equally large disruption comes in the form of elephants chowing down on crops. At night, they will come into fields of nearby settlements for a tasty midnight snack. An elephant's idea of a snack is enough to severely dent your fridge: one thousand Weet-Bix might last an hour, and provide just about enough for a fine aperitif! Farmers aren't famous as the most obliging of hosts, and the elephants seriously try their patience. Everything from throwing stones to fire crackers and gun reports in the air is used to scare them away. Oftentimes, they take offence to the rude disturbance, and that's when things get ugly.

Once again, this is a direct human issue with a straightfor-ward solution- though not one you might expect. Nowadays, farmers are erecting beehives along wires all around their property as a more effective form of defence. When an elephant attempts to cross the border and touches the wire, these beehives resting on it are disturbed. African bees are especially nasty, and they emerge ready to defend their colonies. Now, the greater part of an elephant's skin is as thick as your thumb, and takes more than a paper-cut to hurt it. Hunters opt for high-calibre, heavy-duty weapons termed 'elephant guns,' of the sort that were used in WW1 to punch holes through thick steel plates. So elephants are solidly built, but thick-skinned as they are, elephants certainly aren't thick. They've learnt that bees also go for their ears and trunks which are acutely sensitive. So the ellies stay clear and the farmers keep their crops: as close to a win-win scenario as you could hope for.

My point is that Africa may hold some of the gems of the animal kingdom, but its efforts to protect them are equally admirable. These spaces are truly a hive of innovation and activity, with the numbers to back it up. Sub-Saharan Africa has the 2nd highest output of refugees and the lowest life expectancy of any world region, but it scores slap bang in the middle of rankings for species survival. That's certainly some food for thought.

For those of us not on the front line of conservation, a lot of our knowledge of natural history comes from television. This is where we first meet the majestic and deadly African megafauna, where we tour the world's ecosystems and revel in their wonder. For decades, it was the BBC Natural History Unit at the heart of this, and their champion David Attenborough who quite rightly is revered as a hero in the field. But just how great an effect do blue-chip documentaries really have on saving wildlife? Let's look at a recent example for insight: Planet Earth 2, which boasted a record-breaking 12 million viewers per episode when it aired in 2016.

Immediately there is controversy. In factual TV there always is. Viewers were enraged to learn that the gripping sequence of a young iguana's dash to safety was cut from many takes of similar events. Or similarly that many of the sound effects were produced in studio with Foley, added on in editing. This outcry is entirely unreasonable- the natural environment makes it impossible to do otherwise. The truthfulness of the story is in the hands of the producer, and in almost every case the narrative's crafted to shows genuine natural behaviours. What is interesting about these claims is that viewers want to trust these shows wholeheartedly as a window into the world's environments. Sometimes that can be dangerous.

When was the last time you saw a rhino shot on the likes of <u>Planet Earth</u>, <u>Africa</u>, or (dare I show my age) <u>The Life of Mammals</u>? People like drama of course, but they also want to feel good after watching these shows. Funnily enough, the slaughter of rhinos rarely does that for folks. There is a need to show nature without all the ugly human interference, leading people to underestimate our impact on ecosystems. More recently, one episode per series or perhaps a closing segment has been devoted to these issues, but the problem stands that what viewers see for most of the show is unrealistic, pristine, bustling environments with the events of four years compressed into the space of forty minutes. <u>Planet Earth 2</u> had a full 6% of its script devoted to conservation themes. That's a drastic improvement from its predecessors, but there's still a gap to be filled.

"Surely the impact of documentaries lies in connecting people with animals?" Did I mention I'm a mind-reader? And you're right for the most part- gold stars all round! Each of the episodes of <u>Planet Earth 2</u> raised the profile of the species it showed on-screen. This popularity boost equates to roughly that gained from a world day dedicated to the species. How powerful that might be is unclear. The programmes themselves didn't result in large increases in donations to charity, or large upticks in volunteering. They are amazing creations and people love them- but is that all? I'd argue that the impact of nature documentaries can't be measured by these metrics. It lies in changing people's perspectives, educating and inspiring them.

Let me introduce a little concept of mine I call Conservation's 4C's. It goes like this: saving the world is hard to measure. You could count the number of attractive side-kicks you have, dashing photos in newspapers or evil villains defeated. But superheroes aside, most of us have to stick with estimates of animal populations, or of land area protected. Given that we've only got records for less than 1% of creatures, the former of these can be challenging and/ wildly inaccurate. Unless you're taking censuses of sheep, counting even a single animal species is often very hard work. I have vivid recollections of bat counts at a local park, scanning darkened skies for black silhouettes flying out in their thousands and aiming for a total in the right order of magnitude.

As for protected areas, I've already shown a few of the issues with these. How protected is 'protected,' exactly? Where does 'protected' start and 'not harmed' end? The bottom line is: we love data, but nature has more of it than we can manage. So with the 4C's, the emphasis is on the people side of things. This is far easier to measure, but more importantly, it's also tackling the root cause of problems. These are the four intangibles that mean people are appreciating nature and well-equipped to protect it.

The first 'C' is connection. Now, Gerald Durrell is a name

you'll be hearing several times throughout this book, so it's one that's worth remembering. Best known for his eccentric escapades collecting animals for zoos, Durrell is one of the great names in conservation. He wrote more than a dozen books in his life and is the subject of a modern television series. For a naturalist, that's as good as it gets.

In this guise, Durrell went through dozens of animal companions throughout this life, starting with a dog called Roger. Likewise, David Attenborough spent some time with a colony of bush-babies sharing his home, and when away from her study group of chimps Jane Goodall lived with a pooch named Rusty. What we see is that behind every great environmentalist lies an even greater animal, this bond being critical in directing their life choices.

I call it the Durrell Effect, and the sad thing is that it's an increasingly rare phenomenon. A new term: "Nature Deficit Disorder" has entered the medical lingo, and it's easy to see why. Most children spend at most an hour each day outdoors, with a tiny sliver of that in natural places. By way of comparison, people spend at least five times longer connected to the internet each day. Despite efforts to add green to our cities, we live in a world of concrete and asphalt far removed from other environments, allowing the bustle of new digital distractions to keep us holed up indoors. I'm not pointing fingers, but rather pointing it out as a symptom of the problem. We clearly need a lot more Vitamin N.

Next on the list is curiosity. How would you like to discover a new species? After the last chapter, it might not sound impossible, and that's great- you've learnt something. We live in a world of traffic lights: most things are green-lighted as known, and those things we've yet to figure out are so far removed from our experience or so specialized that we feel trapped in our little box of certainty. These are the red lights, where we daren't go further for fear of doing things wrong, for lack of interest, skills, or equipment. Nature is an exception- a yellow light, if you will. Some things we know (not much, admittedly), but a lot of what we don't is the result of not giving nature the time and space to be observed. That means anyone, anywhere in the world can make a discovery, provided just a little impetus to get them going.

The key to this is wonder (it doesn't begin with C unfortu- nately). Kids are born fascinated by nature, so our job is in keeping the fascination alive. Less than 1 in 5 species has been discovered by science- that's quite incredible if you think about it. There are ant species discovered by picking through frog vomit, and one new lizard species was recently found being served up in a Vietnamese restaurant. You might encounter hundreds of new species in your lifetime, if only you had the presence of mind to notice them.

Another tack is to look at incredible animals. Tardigrades, or water bears, can survive temperatures between -200 degrees Celsius (like Neptune's upper atmosphere) and 150 degrees Celsius (e.g your kitchen oven). They endure through lethal radiation, boiling liquids, pressure equivalent to a pistol's firing barrel, and the vacuum of space. A recent paper found if a nearby star exploded and wiped out the rest of life on Earth, a group of tardigrades living at the bottom of the Mariana Trench would still survive. A collision similar to Pluto colliding with Earth would be necessary to spell the end of them. Are you curious yet?

The third 'C' is creativity, which is where the solutions side of things kicks in. With the first two 'C's in hand, it's time for the fun part- thinking up crazy schemes and ambitious plans to help nature out in a meaningful way. Thriving should always be the goal, not just getting animals to survive. There's a whole chapter devoted to this (number 7), so I'll skip the spoilers, but a large part of the art is in making it fun. Since 1983, a German medical dictionary called Pschyrembel has contained an entry for the Steinlaus, or stone louse. This is a fictitious creature described as a microscopic rodent-like mite, with an insatiable appetite for concrete, bricks, and iron girders. Aside from consuming 28kg of these materials daily, the Steinlaus is apparently known for its medicinal usage in treating bladder, gall and kidney stones. The humorous entry was almost removed by the editors in subsequent editions, but protests from readers saw the Steinlaus retained. An expanded entry even acknowledged its role in the fall of the Berlin Wall. The whole thing is utter nonsense of course, but by having comedy meet with conservation (the Steinlaus is near-extinct) we're able to depoliticize the whole bonanza and make it something that everyone can relate to.

The last of all the 'C's is collaboration. Essentially, this is where things fly or fall. A single swallow certainly couldn't carry a coconut on its own, but if it were strung between two dozen they might have a chance (hmmmm... popular culture analogy?). The same logic applies for environmental challenges. In the year 2000, amidst the hullabaloo of the new millennium, the Earth Charter was released. The charter itself had taken two dozen people six years to write and consisted of an exceedingly concise 2,400 words. The result of all this labour was a powerful framework for future development: 4 pillars, 16 principles, and some 50,000 endorsements to date representing hundreds of millions of people. The ideas it contained weren't particularly revolutionary. Its success was in having such a volume of people commit to its principles, paving the way for sustainable development.

The second example that comes to mind is the recent and ongoing School Strikes 4 Climate. Started by one fifteen-year-old by the name of Greta Thunberg, these events have now involved several million youth worldwide (and counting). They've made some serious waves in the media, and are a major factor in the current widespread awareness of climate change. Youth are great (speaking from experience). They're idealistic, high-level thinkers with hope and a future to fight for. The success of the strikes is proof of that, and of the age-old principle of strength in numbers. Good old Greta.

So that's that. We're almost at the end of the chapter, and what better way to round things out than with a long-winded metaphor? Hang on there, because this one's pure genius once it's explained. It starts like this: nature is a piece of silly putty. You may remember this came up in chapter 1. If any of you poor readers are oblivious to its wonders, silly putty is the semi-liquid slimy stuff that's sold in eggs regardless of whether it's Easter or New Year and is loved by kids universally. What makes silly putty special is a set of rather unusual properties unique to the material, which, as it so happens, are a perfect mirror of nature itself.

To start off, silly putty is an accident. It was created during World War II when the US government set chemists the challenge of creating a synthetic rubber substitute. The stocks of this natural material were dwindling fast, and so scientists across the nation set to work to help the war effort by finding a route for mass-production. Several individuals claim to have invented silly putty around this time, but regardless of who was responsible, the product just didn't pass muster for use in war. It had a nasty tendency to melt and stubbornly refused to hold a solid shape. Then a toy store owner got wind of it, and the rest is history. 4500 tonnes of the stuff have been sold thus far, making it one of the top-selling toys of all time. I'm fiddling with some now as I write!

Nature likewise is somewhat of an accident. We don't know how life started on Earth, or how often it arises on other habitable planets. Speculation is: it's a matter of chance, but like silly putty, it's almost certainly been 'invented' many times over. After that, evolution proceeds in its own sweet time, making slight improvements and testing out thousands of small changes to the basic formula. This is similar to the trial and error of a chemist concocting a range of possible rubber substitutes, varying one small element of the recipe at a time. Nothing much is planned yet still the adaptations of animals improve steadily over time.

Nature is also a timeless classic in a sense, but just like silly putty it goes in and out of favour. The salesman who first put silly putty on the market went heavily into debt before he found success. And, as with nature, originally adults took the greatest interest in it before it was rebranded for kid consumption and youth got hooked. Silly putty even found its way to the moon on the Apollo 8 mission, joining nature in a long and illustrious history of space exploration. To give you a glimpse, fruit flies were the first Earthlings in space, two tortoises were the first creatures to orbit the moon, dogs beat men in reaching orbital altitude and a cockroach was first to be conceived in space. Astronauts have yet to match the last one.

This is interesting enough in itself, but the point of the comparison is not to dwell on the past but rather to illustrate the present. The first thing I tried with my bit of silly putty was to cut it. This represents the bulldozer moment in any nature documentary: when the nasty corporate powers knock down a forest and you see the orangutans in tears. When we remove habitats, we cut the putty. If the land is cleared, you throw away the cut pieces. If you built a road through the middle of it, then you just end up with two halves. Fragmented habitats aren't nearly as good as connected ones, so both these actions cause harm.

Next, in times of hardship when human society is placed under pressure, we do what we can to vent the pain and squeeze nature. Silly putty is easy enough to mould under this gradual pressure, and that's just what we're seeing happen in today's world. Examples might be the hunting of wild animals for bushmeat or creatures caught in the crossfire in times of war. It might be overfishing, or insecticides decimating insect populations, or the illegal wildlife trade for pets and ivory.

In times of plenty, when many people can afford to purchase small luxuries from across the world, we do what any kid does upon receiving such a toy. We stretch it. If the change to affluence is slow, then the process can continue a little while. If the change is fast and resources are unsustainably harvested, then in places nature will snap. If you pull silly putty at speed, it fractures, however try with a steady tug and it will stretch out as fine as a strand of hair.

What's inspiring about this comparison is that, like putty, nature is tough. One of the substance's extraordinary properties is that the more force you apply to it, the tougher it becomes. You can hit it all you like with a hammer without making the slightest dent. So in spite of all the damage we continue to cause, some animals will always manage to adapt in new ways to make the most of their changing environment. An octopus off Florida's coastline has been filmed stealing a quick fish snack from fishermen's traps. Some herons have learnt to fish using bread from human tourists as bait. Crows are enlisting cars to crush their nuts, young gorillas in the Congo have learnt to dismantle hunters' snares, and peregrine falcons have found both 5-star balconies and all-you-can-eat pigeon buffets in the form of skyscrapers.

I've got a science lab in my basement, so I added small lumps of silly putty to a range of organic solvents. These represent nasty chemical pollutants being churned out of factories, and-lo and behold- the putty added to them promptly dissolved. Think of the grisly pictures of seabirds' stomachs stuffed with plastic or the toxic sewer sludge leaking into creeks and rivers. Nature and artificial chemical cocktails just aren't a great combo. That said, once this stuff gets onto your clothes, it's an absolute nightmare to remove. So as much as we try to keep nature away, it'll always be around us. In the most inhospitable, people-infested city centres there are still animals to be found. We just forget to pay attention and they slip beneath our gaze.

When silly putty was first sold, an egg contained 1 ounce (30 grams) of the stuff. Nowadays, that same container for the same price holds less than half that. In a similar span, since 1970, the populations of animals in the wild have dropped by 60% or more. The two things really are a perfect match. When you heat silly putty, it becomes runnier and easier to shape. With climate change, animals are forced to relocate, and so all our human impacts on them become exacerbated. But the really important point to note is that left alone, silly putty spreads out like a liquid and claims all available space. If we remove our negative impact on wildlife, then it really can reclaim lost ground, restore areas to a natural state and rebound. In case you're wondering, silly putty bounces. Nature might.

This chapter might seem negative, but there needs to be a problem for a solution to arise. My hope is that you now have a better understanding of some of the challenges nature faces, and just a glimmer of hope that if we work tactically we can help fix them. So, with that mind-boggling putty analogy aside, it's time to move on to greater things: no less than intelligence itself. It's time to switch on your brain and open your mind. Onwards!

4

Great Minds

Why is a raven like a writing desk? This infamous riddle, posed by Lewis Carroll in his bestselling Alice's Adventures in Wonderland, was never intended to be answered. Yet despite this, the question has plagued intellectuals for the past 150 years. There have certainly been some valiant attempts at solutions. "Because it bodes ill for owed bills" is neat, as is the more direct response "Because Poe wrote on both." You may prefer: "Because a writing desk is a rest for pens and a raven is a pest for wrens," or "Because the notes for which they are noted are not noted for being musical notes." My favourite is the absurdist response: "Because there's a b in both, and because there's an n in neither." I'll leave you to figure that one out.

We've made it this far in the book without any serious challenge to the concept of 'animals' as distinct to 'humans.' That's about to change. This chapter I'll use to ease you into it, exploring the more remarkable traits of the animal world. The one that follows is intended to debunk any arguments that might be used to plea a special case for humanity. Alongside both, we'll dip our toes into the quagmire of intelligence and try to puzzle out exactly how it's measured. Are you ready?

Aesop was a Greek storyteller, most famous for a collection of his works known by the uninspired name of <u>Aesop's Fables</u>. Although none of his original writings survive, a large number of tales credited to him have none-the-less been collected from different cultures over the years, giving rise to this remarkable work of literature. One of the tales, titled "The Crow and the Pitcher," describes some of the most remarkable evidence of bird intelligence to date. It goes as follows...

"A crow, ready to die of thirst, flew with joy to a pitcher hoping to find some water in it. He found some there, to be sure, but only a little drop at the bottom which he was quite unable to reach. He then tried to overturn the pitcher, but it was too heavy. So he gathered up some pebbles, with which the ground was covered, and, taking them one by one in his beak, dropped them into the pitcher. By this means the water gradually reached the top, and he was able to drink at his ease."

The moral of the story, quoted below it, is that "Necessity is the mother of invention." By displacing the water with pebbles, the crow in the tale shows an intelligence greater than that of a young human child. What's more, this story has been replicated many times by researchers in the modern world. One crow, known as 007, made history on UK national television when he completed an 8-step puzzle in a matter of minutes. And I still struggle with child-proof bottles!

Those who study cognition use a measure called the en-cephalization quotient as a proxy for intelligence. In simple terms, this is the size of the brain relative to the animal's body mass. Generally, fish score lowest, then birds, then mammals, and finally primates. Crows and related species (the corvids) sit just below primates on this scale. That means they've got very big brains. Consider some of these wonderful anecdotes:

Many years back in the US state of Montana, a crow devised one of the most fiendish school pranks ever attempted. This cunning corvid went round all the homes in the local neighbourhood with a seducing cry: "Here boy! Here boy!" accompanied by the occasional whistle for urgency. At each stop, one more dog heeded the summons, and soon a pack of them had formed behind the crow. Eventually, it flapped over onto the branch of an oak tree beside the oval of a local university. Sure enough, the expectant dogs gathered around and waited patiently. When the end of class bell chimed, all hell broke loose. The crow flew low through the mob of students emerging, followed by the pack of dogs hot on its heels. I only wish I could have been there to see the result.

These birds, of course, are famous for their ability to mimic human speech. They've been known to talk in French, German and Russian as well as English- and apparently Latin as well. Augustus Caesar was returning home from his crowning victory at the Battle of Actium when he passed a man in the street clutching a raven. With apparent spontaneity, the bird called out "Greetings to Caesar, our victorious commander." Augustus was delighted and paid twenty thousand sesterces for the bird, equivalent to roughly 50,000 USD today. That's one valuable bird that assured itself a comfy palace life.

The slur "bird-brained" arose several centuries ago upon discovering that birds lacked a part of the brain known as the neocortex. This takes up most of the space inside our skulls and forms a thick outer layer for the brain. It's also considered the seat of mammalian intelligence, as opposed to the primitive 'reptilian' part of the brain that it surrounds. Since birds lacked this key component, it was assumed that they were incapable of higher-level thinking. The preconception stood largely unchallenged until one decade back, when some enterprising scientists showed near-identical structures in a completely different part of birds' brains. Why didn't they look before? That's science for you.

Birds share other similarities with the mammalian brain. The amigdola, controlling emotions, and hippocampus, controlling spatial recognition, combine to allow complex memories to form for birds as with humans. Plus they have one trick humans lack: the ability to separate the two hemispheres of their brain, allowing one half to sleep whilst the other carries on functioning. Now that's a superpower I wish I had!

Crows themselves are much more alike to us than we give them credit for. Both humans and crows take an unhealthy interest in caffeine and cigarettes, both are good at recognizing faces, and both possess a cruel sense of humour. One wonderful tale is that of a Swedish woman who took to feeding magpies. Each morning, she would throw out food scraps for them in the yard. Each morning, the magpies would come and eat them. When she withdrew into the house, the magpies would fly towards the window. The woman was impressed by this and gave them extra food. Whenever she walked past a window, the magpies would tap on it with their beaks, gaining a further reward. One day, the doorbell rang, and the woman answered it. No-one was there. It happened again, and again until finally she caught the culprit in the act. It was her magpies, of course.

The woman's husband grew tired of their demanding attention and would make gestures as if to throw objects at them as he passed. The offended magpies promptly vented their feelings by crapping all over his car windshield on the driver's side. They did this regularly, every morning. He switched the places of his and his wife's car, but the vandals were not fooled and continued to make their statement. That's people 0, nature 1.

The important thing to recognize here is that animals are individuals, with personalities, self-awareness, emotions, and consciousness. Just as Mandela and Michael Jackson weren't exactly twins, so too will one crow be very much distinct from every other. Only they're better than us at spotting the differences between individuals of other species, hence their capacity for revenge. Scientists working with crows know this and take care to wear disguises if they intend to do anything offensive to the birds. The thing is, crows don't look much like us at all, hence it's difficult for people to accept them as intelligent. A much more familiar form is our close relative the chimpanzee.

Meet Natasha, a rescued chimp living in Uganda. Natasha first drew notice to herself when she started clapping her hands during feeding times at the Ngamba Island sanctuary. The keepers, confused, were invariably drawn over to her, and she ended up with the lion's share of the food. As time went on, she learnt other tricks: luring in visitors until they were close enough to be splashed with water, and testing her enclosure's electric fence by throwing sticks at it. This devious method allowed her to make multiple escapes. When a study of 106 chimps in sanctuaries globally was performed, Natasha topped the intelligence rankings by a fair margin. In doing so, she earned herself the title of 'Natasha Einstein, the chimpanzee valedictorian.' Clearly, she was the crème de la crème of her species, but not without some stiff competition.

One possible contender is Ayumu. Born in Kyoto University's

Primate Research Institute in Japan, Ayumu made headlines when, aged 7, he proved himself to have a working memory better than any human. Ayumu lives in a group of 14 chimps, enjoying his life in a recreated West African rainforest the centre has set up as their home. Each day, he emerges from the greenery at the appointed time and presents himself for testing at a simple game the researchers have devised. The numbers 1 to 9 flash onto a screen, appearing in randomized locations. Once Ayumu presses the number 1, the others disappear behind white squares. He then has to press each number's box in ascending order.

The researchers have also trained up a group of university students to compete with the chimps. Suffice to say, they don't stand a chance. Ayumu can memorize all the numbers in 60 milliseconds- that's the time it takes you to blink- and gets them right over 90% of the time. The best human students take about half a second to memorize the digits, and can't hope to match the chimp for precision. But the real test came when Ayumu faced up against three-time world memory champion Ben Pridmore. Pridmore can memorize the order of a shuffled deck of cards in 25 seconds, yet still scored three times worse than the chimp prodigy when they competed live on television. That's great for Ayumu, but seriously embarrassing for humanity.

Do you think you can do better? Want to restore the good name of Homo sapiens? The next page has a printed version of the test for you. When you're ready, turn over and take a glance at the spread, then cover the numbers and use the boxes below to fill them in. Or, if you'd rather the full experience, look up elliotconnor.com/projects for the digital version.

* * *

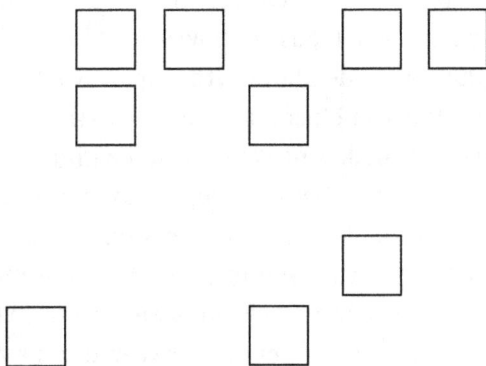

9 3   6 5

2   7

4

8   1

* * *

How did you go? Chances are you cheated or failed- either way, Ayumu wins. Impressive, right? You might argue that's just one test- it only shows chimps are good at memorizing, like that friend who remembers everyone's phone numbers. Any old fool can do that. Part of the problem is that sophisticated reasoning cannot be deduced through observation.

If I were to watch the great Greek philosopher Plato musing on the steps of the Parthenon, I might naturally assume out of context that he was depressed, or that he had some sort of social phobia. German biologist Jakob von Uexküll described this in what he called an 'umwelt'- a perception of one's surroundings unique to each living thing. A panda's view of the world would be very different from Plato's, just as it would be different from a pigeon's. Simply looking at any one of these three cannot possibly tell me what they are thinking.

Umwelt theory is powerful in explaining the limitations of our understanding other lifeforms. Chickens, horses, dogs, honeybees, spiders, salamanders, fish, crows, lions, bears, macaques, chimps and toads all show signs of numeracy. Does that mean that can understand Pythagoras? No! Should they? Of course not! It doesn't factor into their lives (nor does it ours) on a daily basis. It makes no sense to add within the frame of their umvelt, which is better occupied by more serviceable skills. We use basic behavioural tests on animals because they're easy to construct and interpret. That's fine, but we need to be cautious of the conclusions we draw from them and keep in mind the vast differences between our umvelts. The classic example of this is the mirror test.

The theory goes, if you paint a spot on an animal's forehead and place it in front of a mirror, any intelligent animal with a sense of self would reach out and touch the strange mark. This shows that not only do they recognize themselves in the mirror, but they also have some sort of mental self-image that allows them to puzzle over the unfamiliar spot on their forehead. These tests have been tried on Asian elephants, bottlenose dolphins, orcas, magpies, great apes such as chimps, bonobos, orangutans and gorillas, and even ants with positive results. Then in 2018, a team came along claiming that a fish- the bluestreak cleaner wrasse- had also passed. People weren't happy about that.

The scientists had taken 10 wild wrasses and placed them in front of the mirrors. Initially, they would react aggressively towards their reflection, believing it to be a different fish and therefore a rival. Soon, though, their behaviour changed. They would swim upside down, or mock charge the mirror but pull up short. These sorts of moves had never been witnessed before, suggesting the fish were directly interacting with their reflections. Put a caveman in front of a mirror, and they'd likely do a similar thing.

The team of scientists then injected a harmless gel under these wrasses' skins, coloured either brown or transparent. Those with the coloured gel were observed trying to scrape it off, whereas the transparent test cases made no such attempts. The scientist watching the tests was so surprised he fell back-wards off his chair. Nobody expected this.

Now, great apes, cetaceans (dolphins and whales), elephants and corvids are all considered highly intelligent animals. Ants are shown some respect for the brains they pack into their size. But fishes? Far from it. The published results weren't taken lightly. Some were impressed and praised the team. Others like Gordan Gallup- who invented the mirror test- sought to discredit the findings. He claimed the behaviour was likely observed as a by-product of the fish's habits feeding off other fishes' parasites. Hence their interest in the blemish. Primate researcher Frans de Waal suggests it might be valid, but rather interpreted on a continuum of self-awareness as opposed to a black-and-white model. Fishes, he says, show some evidence, but that for apes and dolphins and crows is still substantially stronger.

What the incident does show conclusively is that tests like this have major flaws. Fishes like the cleaner wrasse are seriously disadvantaged because they don't have arms or trunks to reach out and touch the mark. They can't even be given the painted spot like other animals, because any non-toxic paint would wash off them underwater. Dye injections are the alternative, but far less convenient for research. Large mirrors have been set up in the Amazon and filmed, revealing all sorts of animals interacting with them. These behaviours, however, are up for interpretation and certainly don't count as passing the mirror test. Brilliance, as they ought to say, is in the eye of the beholder.

But back to brains. I mentioned earlier the concept of the encephalization quotient- EQ for short. This measures how large an animal's brain is relative to what's expected for its size. Humans score around 6 or 7, bottlenose dolphins get over 4, chimps and ravens both sit at roughly 2.5, elephants at 2, sealions at 1.5, horses and cats at 1, mice and hippos at 0.5. It all seems very straightforward, but being science, it is anything but.

Take dinosaurs- I've done no small amount of research into these 'dim-witted beasts'. As it turns out, diplodocus had an EQ of about 0.6, triceratops at 0.7, iguanodons at 2.5. That last one's a surprise, but it gets better yet. Velociraptors come out with an EQ of 5.5 and Tyrannosaurus rex tops the scale as a dead heat with humans: 6.5. Think about that for a second.

Isn't that incredible! If it weren't for that blasted meteor, Tyrannosaurs might be sitting at computers typing out emails today, or counting the number of stars in the universe. Right?

EQ was only ever designed as a measure of intelligence between related species. Its measurement is subtly different for reptiles like dinosaurs, so comparing them to humans is dubious to say the least. A better comparison might be to say that early human species (hominids) were of comparable intelligence to dolphins. That's true according to their EQ values. Women's brains are on average 14% smaller than men's, but their body size is also typically smaller. For the EQ ratio, it all levels out more or less. Then again, women have been shown to use their brains more efficiently than men, so in the intelligence race it's looking like blokes are running a few steps behind.

They say that actions speak louder than words. When it comes to cross-species understanding, that's more true than ever. There's no common language we know of between humans and apes, so gesture is used to fill the gap. Koko the gorilla was born in San Francisco zoo and, from the age of 1, worked with researcher Francine Patterson who set about teaching her the rudiments of American sign language. What began as a PhD rapidly became a life's work, and, founding The Gorilla Foundation, Francine moved Koko into a more spacious abode in the Santa Cruz Mountains. There, until her death in 2018, Koko worked and lived alongside Francine. She learnt over 1,000 separate signs, and understood 2,000 words of spoken English. At one point, Koko requested a pet cat, and rapidly became attached to the kitten she was given. She called it 'All Ball,' and played the part of its loving mother. When actor Robin Williams paid a visit, she asked specially for a tickle and promptly stole his glasses.

Kanzi the bonobo is another example. Bonobos are both the gentlest and least good-looking of the great apes. Half of that I reckon we inherited from them, being our closest relatives. Confined to a small range in the Congo rainforest and displaying (so I'm told) sexual behaviour startlingly similar to humans, it's no great surprise that bonobos rarely feature in zoos. As a species, they keep largely to themselves, with a very low profile in the human world. Except one.

Kanzi was raised in a sanctuary in Georgia, where his mother was part of an attempt to teach language to apes. The lessons used a system of abstract symbols called lexigrams, and though his mother struggled, Kanzi sat in on classes and took well to it. With a similar vocabulary to Koko and an even greater understanding of the spoken word, Kanzi has since shown himself adept at a whole range of other 'human' pursuits. From starting fires and roasting marshmallows to creating stone tools and writing- even playing the piano- it's hard to find something that doesn't interest this curious ape. From the close relationship Kanzi shares with the researchers and fellow apes, it's becoming increasingly clear as well that bonobos use their own high-pitched language to communicate. This is a language we have yet to understand, but which could hold the key to an even closer partnership.

We're at the point in the piece where I'm meant to make some comment like: "So you see, bonobos have learnt human language but we're too dim-witted to learn theirs." It's an interesting thought, but little more than that. Sadly bonobos have shown themselves thus far disinclined to teach us or to create special institutions in which to do so. There's a common theme we're seeing though in that humans have to work with animals to understand more about them. Often we take the role of instructors, leading animals through the paces as they learn various skills or testing routines. Sometimes the animals' actions are purely spontaneous, driven by a powerful inner intellect that leaves the researcher as a fascinated bystander. Is there ever a situation when animals end up playing the teacher?

Let's try something new and opt for a running tally. Who invented electricity? Animals. Medicine? Animals. Democracy? Animals. What about GPS, radar, public toilets, music, art & architecture, lift-sharing, city planning, air-conditioning, chemical explosives, surfing and perfume? Animals. First into space? First on a hot-air balloon? Animals and animals. The list goes on and on, but for the sake of retaining some scrap of human dignity, I'll stop there. That's animals 17, humans nil.

Okay, so these last two were admittedly human-engineered, and the animals in question showed little skill in piloting. But go back to the start of that list and read it again. We certainly didn't give animals electricity. We didn't teach them to use medicine or give them GPS instruction manuals. These are inventions that animals themselves created, and humans in one way or another have copied from. Fortunately for us, there's no such thing as plagiarism in the survival of the fittest. Otherwise, we'd be under a lot of legal scrutiny.

The word for this phenomenon is 'biomimicry': humans taking nature's blueprints and redrawing them for our own little commodities and systems. Through this process, we have or will soon have phone screens inspired by seashells, bullet-trains inspired by kingfishers, antifreeze inspired by cod, painless surgical needles inspired by mosquitoes, wetsuits inspired by otters, wind-turbines inspired by whales, space rovers inspired by desert spiders, gene-editing technologies inspired by bacteria and many, many more wonderful things. I mentioned de-extinction a little earlier in the piece, and it turns out we'd be nowhere without microbes to guide us in the method and techniques for this. The same applies for genetically modified foods, meaning most of the cereal, bread and biscuits that you eat.

Aristotle in his Politics said "Man is by nature a political animal." He was more right than he knew. The democracy that defined his home city of ancient Athens is actually not uncommon across the animal kingdom. Buffalos use it, as do wild dogs, chimpanzees, pigeons, ants, and possibly cockroaches. The most impressive example, however, can be seen in bees.

When a bee colony wants to relocate, they face a challenge. With 10,000 bees and no hierarchy, they're a little like a Parliament with too many politicians. Except that bees beat politicans hands down! The first step is to send out 300 of the most experienced worker bees to scout out the surrounding area and shortlist some of the best spots. Once a scout returns with a site in mind, she performs what's playfully termed a "waggle dance." Amidst lots of complex footwork that would put any tap dancer to shame, the scout signals not only the location of the site but its suitability for the colony. All this is coded into the routine. The longer she dances, and the more times she repeats the steps, the better her rating. A bit like TripAdvisor, but a whole lot more entertaining.

Once most of the scouts have returned and done their little dances, a sort of debate takes place whereby every bee in the colony dances the dance of their favourite option. Soon enough, one choice wins out and all the bees are dancing this. The scouts then muster up the entire colony and lead them to their new home, no doubt hoping for a 5-star review and a tip for service.

Being 98% composed of females, it's hardly a surprise that bee society does so well at politics. If only every nation had a Jacinta Ardern at their head! The 'queen bee' title is also somewhat misleading, suggesting more power than she has. The queen bee is nothing more than the mother of the colony and a marvellously efficient egg-layer. She does little but sit around all day and spread her influence via lots of public servants. That's a very modern monarchy!

We're learning to make bendy batteries thanks to electric eels, picking up architecture tips from termites, learning explosive chemistry from bombardier beetles, and plenty more just because some of us have started looking. You probably didn't know that monkeys self-medicate, but more pressing by far is the fact we've got anywhere between half and two-thirds of our pharmaceuticals from plants. Taking inspiration directly from nature doesn't mean we have to branch out. Often nature provides solutions prepackaged and just waiting to be found, as opposed to the long and arduous process of humans using trial and error.

This isn't the genius of a species or even an animal, but of evolution itself. For the record, humans are not "more evolved," and I reserve the right to shoot anyone that says so. Evolution is a process of gradual improvement, and we've all been running as long as each other in this race. Biodiversity is life's treasure trove, but to us it's also an instruction manual. As for the rest of the animals vs. humans tally, I'll leave you the homework of figuring it out. But believe you me: if we all took a good long look at nature, we'd be realizing Hollywood's sci-fi tech imaginings within my lifetime.

How about immortality? A cure for cancer? Better quantum computers? These are a few of the holy grails of our time. And nature plays the patient lecturer to humanity its unruly student. Immortal jellyfish don't have the most imaginative of names, but having cracked the secret of reverse-aging they certainly stand true to this claim. The same applies for certain flatworms and even tiny ocean krill to the extent that they can turn back puberty.

Telomeres are the little clocks in our cells, responsible for all the signs we know as ageing. Short telomeres mean you're going granny ways, whilst long telomeres are proof of youthful vigour. We've got a drug called telomerase that's starting to allow us to fiddle around with these nifty little machines. Sadly it's not as easy as resetting clock-hands, but give science a decade or two and we might just have ourselves micromachines small enough to get inside our cells and make the necessary adjustments. By looking at 'immortal' animals, we can work out all the implications these changes may have. After all, we need the right settings to use for our body's own time machines. We can make sure we stay looking young and beautiful without the nasty side-effects. Oh... and yes, you would be able to choose your own age (dibs eighteen!).

Then there's cancer, that most pervasive of human diseases. Half of us will get it in our lifetime, making us exceptionally vulnerable amongst animals. For most species, the chance is maybe 20-40%. The difference is that we stick around a while, far beyond our natural life expectancy. Elephants are equally long-lived with a 70-year average lifespan. They also have 100 times as many cells as us to start with. More cells mean more chance of nasty mutations and cancer, so elephants should be dropping like flies from the disease. Yet there are hardly any cases of elephant cancer recorded. Scientists stumbled upon this fact recently, and the reason they found is that elephants have evolved to fight it. They've got 20 times as many copies of a cancer-suppressing gene called TP53 than do us measly humans. What's more, they've come up with a clever trick for damaged cells to simply shut down and not clone like crazy. No replication of heywire cells means no cancer.

Squirrels, horses, whales, bats, and mole rats all have been observed to share this near-immunity, and the methods they use to achieve this are just as diverse and unexplored. Imagine if one day in the future when you're working on your designer baby, you're adding in a pinch of elephant genes to keep them cancer-free? Come to think of it, a trunk would be rather handy. Gene-editing technology from bacteria and the knowledge to use it from elephants together might be able to cure our deadliest disease. Who'd have thought?

By now, you're probably thinking I'm either mad or the most amazing futurist ever. I'll admit more likely to be the former. The monkey-computer bit is still to come (time-travel's confusing, isn't it?), so perhaps you're a little sceptical that animals could teach us any computer skills. That's understandable. Kanzi the bonobo might be a whiz with Skype (trust me, he is), but quantum computing is another world entirely. Even for the less down-to-earth folks amongst us, surely that's ridiculous?

A bloke called Neils Bohr laid the foundation of quantum theory and won a Nobel Prize for his troubles. So the fact that he said paradoxes were needed for progress ought to give you some indication of how strange this subject is. Get ready for some very long-winded analogies. Quantum is all about the actions of tiny particles smaller than atoms which ignore our classical science and stretch our understanding of what's possible. Time to dig out that jumbo-sized magnifying glass and dive right in!

First, there's my daily commute. This commute, however, is of the less-than-ordinary kind. I do the same commute every day via a dozen different routes: by train, by bus, on foot, motorbike, sidecar, limousine, helicopter, paragliding, skateboarding and pogo-stick jumping to name but a few. All of these take place at once, mind you, so I always get to school in the shortest possible time. That's what tiny light particles called photons do when they reach plants that use them for photosynthesis. By trying out multiple routes simultaneously, the photons always get where they're going as fast as possible. It's clever, isn't it? The quickest possible route means a minimum of energy loss and a maximum of nice green veg for your dinner table. Sure it sounds impossible, but that's quantum for you.

Or take my maths lessons, where I unnerve my classmates by walking through the back wall into my next class in the adjoining room. That's what chemicals called enzymes do in all living creatures to help perform the life-giving chemical reactions they need more quickly. Ghosts are real and they're inside your body, because these enzymes teleport material from outside of cells into them right past solid walls.

Finally my school formal last year, spinning with that special someone on the dance floor. I'm no dancer but us two were inseparable, reading each others' moves before we'd made them. That's akin to the nigh-inexplicable interactions of special particles that, we think, might give birds their incredible sense of direction from Earth's magnetic field. I didn't mention this at the time, which was probably for the best. These particles do a sort of dance in the birds' brains, allowing them to literally see magnetic fields around them. And if this particle pair were separated a universe apart, they'd still mirror each other's moves just the same. Spooky.

That wasn't so hard, now, was it? Though I doubt it's all that much clearer as a result. As soon as we can better observe and begin to explain these sorts of effects in animals, we've got a whole new toolbox at our disposal. That's the point at which we can start doing funky stuff with the weirdness ourselves. What might come out of that is hard to know, but it will certainly be spectacular.

We started this chapter with a perplexing riddle. Now we're at the perplexing science stage of things. Remember, this chapter is about showing you how wonderfully brilliant animals are, and how thick humans can be in comparison. So it only makes sense to end on everything we don't know about nature. This is the IQ test where there are no winners.

In 2010, London's Natural History Museum opened the doors to a new building wherein was what they called an "Identification & Advisory Service." Everyone else immediately came to know it as the "British X-Files." The idea is that anyone can bring in mysterious animals or other finds they come across to be identified by the museum's team of experts. People may be seeking advice about how to deal with the strange find, or sometimes claiming that they've chanced upon a new species. In the time since it opened, the centre has been inundated with visitors and requests nationwide. Over 10,000 inquiries each year are directed their way, or about two-dozen each day.

Most of the time, their small group of staff can handle the questions with little difficulty. In rare cases, they'll pass on the specimen to one of the 300 researchers in the museum for a second opinion. Through these means, they're shown a 'sabre-tooth cat skull' to be that of a Chinese water deer, 'dragon skulls' to be the pelvis bones of seabirds, and dozens of 'meteorite fragments' to be old balls of kitchen foil. Other finds are not so easy. Like the mysterious 'star jelly' that appeared in a nature reserve at about the same time a meteor crashed to earth in Russia. Fantacists believed it to be a sort of alien life- a 'space slime,' if you will. The museum's identification team weren't so quick to jump to conclusions. DNA analysis was performed and showed bits of magpie and frog DNA in the transparent goo. Speculations were made as to the sequence of events behind this strange combination, but the real backstory remains unknown.

There are other, more basic phenomena we can't yet explain. You've heard the saying 'it's raining cats and dogs.' To date, I'm yet to hear of these domestics dropping from the heavens, but there are plenty of eyewitness accounts for everything from falling frogs to spiders, tadpoles, fish, eels, worms and snakes tumbling from the sky. Some claim that tornadoes or waterspouts pick up the creatures from their habitat and drop them miles away. For single animals being dropped, birds of prey are often the cause, but in large numbers this seems unfeasible. No conclusive evidence to support either mechanism has ever been found, and mystery remains unsolved as all great mysteries should.

Another apparently miracle is the self-healing powers of sponges. Some of the simplest organisms on this planet, sponges are the living, blobby, marine counterparts of your yellow kitchen cleaning implement. They spend their lives sitting at the bottom of the ocean, filtering water for the nutrients it carries. Not a tremendously exciting existence in my books. Sponges do, however, have one trick that you couldn't in your wildest dreams hope to copy. If you put a sponge in a blender or force it through a very fine sieve until only its individual cells remain, it can reassemble itself from scratch and go on living and feeding as it always has. Truly bewildering.

One last puzzle for you to ponder. Remember the false killer whales from last chapter? Well, they have one of the worst names in the animal kingdom. They're not killers (unless you're a fish), they're not whales (they're dolphins), and as for the false part... well, that's just plain rude. Would you call your cat a 'false lion,' or your fork a 'false spoon?' The mystery isn't their name: that's just human stupidity. The mystery is how on dozens of occasions they've been observed deliberately stranding themselves on beaches in large numbers. They're not the only species to do so: sperm whales, beaked whales, pilot whales and all sorts of other dolphins have been observed doing likewise. The largest such mass stranding known was 835 of them, but smaller groups of just a few individuals are also common.

This is a tragic occurrence, and researchers are understandably desperate to discover the cause. The usual long list of suspects has been compiled: shipping noise and military sonar, pollution, pneumonia, magnetic disturbances, diseases and trauma all have a following and varying amounts of evidence behind them. In some cases, we can work out the cause. It may have been a recent shark attack, say, or a ship collision. Far more often, though, we're in the dark as to how this comes about. Even when the animals can be rescued and released back into the sea they seem determined to beach themselves repeatedly. Often a call is made to have them shot as a quick end to this suffering. Aristotle and his contemporaries reported such incidents, as did settlers in Puritan New England. So modern technologies and shipping are likely only one part of the answer, and it may take many years yet to unravel the major causes.

If chapter 2 showed you anything, it was that our under-standing of nature is far from complete. These stories only emphasize the massive knowledge gap. Given that we're speaking of intelligence, though, I feel it only fitting to end on a note towards how unintelligent we are at describing other animals' smarts. One infamous test was used to determine if elephants use tools. The researchers provided a test subject with some food on a ledge out of reach and a branch with which to knock it down. It seems straightforward enough- either the elephant uses the branch or it doesn't. When it failed to do so, the verdict was ruled that elephants aren't intelligent enough to use tools.

What the scientists failed to realize in running this test is that elephants orientate themselves by smell through their trunk. This is their most powerful sense, equivalent to eyesight for us. For the elephant to pick up the branch and use it as required would have been comparable to a human playing marco polo with a banana. Later the experiment was revised and repeated with a stool that the elephant could pull over to stand on. This time, it passed with flying colours and the original conclusion was revoked. You know, sometimes I wonder whose intelligence we're really testing with these studies.

By contrast, the late conservationist Lawrence Anthony witnessed some amazing feats of elephant intelligence simply by befriending them. This was a rogue herd, their matriarch shot, and saved from culling by rapid relocation to Anthony's own reserve. The alternative was a a lot of dead elephants that would have had to have been put down. So it was a very brave move that Anthony made, and one which the elephants didn't make any easier.

These animals were kept in a temporary enclosure for a short time to let them acclimatise to their new surroundings. In due course, two of them toppled a mighty tree, working in tandem to bring it down on a nearby section of the enclosure's electric fence. The power shorted and the elephants escaped. These creatures could sense when the power was down and knew precisely the right moment to perform their act in order to avoid being found out. Elephants can be very quiet when they need to be. In spite of this poor start, Lawrence developed a strong relationship with the herd. They saved him from a bushfire on the reserve, and came to mourn at his house after his death. Both of these are incredible proofs of sophisticated thought and feeling.

Intelligence the way we define it now measures how similar an animal is to humans. Elephants do fairly well in the comparison, as do chimpanzees, crows and dolphins. These are the characters that have come up time and again in this book thus far. Yet if we distance ourselves a little, we can appreciate how flawed that measure really is. It's like that classic child's feat whereby after losing at a game a new rule is invented to mean they always have to win. We set the boundaries such that we always come out on top, and even then there's no shortage of competition.

IQ tests rely on a form of 'general intelligence' presumed to be inherent and applicable to all tasks. Then again, when we look at how varied life is one has to doubt such a concept could apply to more than a handful of similar species. There are too many factors coming into play here to really measure and rank animals on. Abstract thinking, knowledge acquisition, learning and logic, problem-solving, emotion, communication, philosophy, deduction... these are some of the many hallmarks of human intelligence, but why would they apply beyond our species?

Mozart is probably the most famous composer of all time- a musical genius and incidentally, a bird fancier as well. Mendeleev is one of our most famous scientists- causing many hours of high-school torment when we invented the periodic table. I could ask the question: which of these men was the brightest? You may have an opinion, but like comparing apples and oranges they have too many disparate qualities to compare.

Albert Einstein once remarked: "It is not that I'm so smart. But I stay with the questions much longer." So, in keeping with one of the greatest minds of our age, let us end with a question or two for you to ponder. Can we redefine intelligence? Can we create a meaning for the word that is consistent and fair across all species? A simple answer. We'll see...

5

A Monkey's Microwave

There's a famous question that goes: if you give an infinite amount of monkeys an infinite amount of typewriters would they eventually produce the complete works of Shakespeare? As is often the case with such meaningless hypotheticals, many a great mind has devoted themselves to solving it over the years. Douglas Adams thought it merited reference in his Hitchhiker's Guide to the Galaxy, modern philosopher R.G. Collingwood firmly opposed it in his books, and an episode of the Simpsons showed a thousand monkeys making progress on the opening to Dickens' A Tale of Two Cities. That's not a bad start.

The question's origin has been traced back all the way to Aristotle, emerging in its modern form thanks to the musings of the great French mathematician Émile Borel. Behind this thought experiment lies the premise that a monkey given a typewriter would press keys at random for as long as the opportunity was presented. Personally, I'm a little sceptical of that idea. If the last chapter has shown anything, it's that animals are far from random- so what would actually go down?

In 2003, a team of six macaques at Devon zoo in the UK set about this very task. Having been provided with a single keyboard between them and a month in which to work their brilliance, the pressure was on for the monkey team. Writer's block set in, and the resulting five pages of manuscript were filled mostly by the letter 's'. Still, undaunted, they published their work in a limited edition volume entitled <u>Notes Towards the Complete Works of Shakespeare</u>, charging 25 pounds apiece. As I write this, the work has a single review on Google whose five-star rating describes it in flowery terms as "sheer brilliance" and a "must-read for any literary critic or connoisseur."

Little-known masterpiece or almighty hoax? Admittedly I wrote that particular review, feeling that the macaque authors could use some encouragement. The experiment itself does however bring to light some interesting questions. Were the monkeys random in their typing? The answer is a resounding 'no'. Of the 13,000 characters in the work, 75% are the letter 's', 10% are 'q's, 7% are 'a's and 4.5% are 'g's with a total of 16 letters out of 26 represented.

Did they write any words? Strictly speaking, no. They only began to use spaces to separate characters in the final few lines of unintelligible nonsense. That said, I put the work through two popular spell-checkers, which gave 0 and 15 errors respectively. The first <u>Harry Potter</u> book when put through a similar test comes out at about 1000! If you discount the need for spaces, then a scrabble dictionary will provide half a dozen words and initialisms hidden therein: from blood-potassium medication to well-armed submarines, politics and sports jargon, as well as the staple 'aa' referring to a variety of lava flow.

The monkeys clearly took interest in the device, fascinated by the appearance of letters on the screen at the pressing of a key. And naturally, tempers rose on occasion- the dominant male was observed trying to beat the equipment with a stone and others vented their anger by soiling the protective casing. In the end, as the test's chief instigator noted, "it was designed more as a performance than anything else." So what does it show?

The six simian writers: Elmo, Gum, Heather, Holly, Mistletoe and Rowan certainly didn't produce anything in the realm of sonnets or stageplays. Why would they? This question is the start and end of almost every animal intelligence debate, because how can we measure other creatures by human standards and expect them to compare? These letters they typed out are meaningless to them- it's not their language. I'd be writing gobbledegook if you made me type in Russian, but thankfully I don't have to.

It's a pertinent point, but not a great one- resulting in more of a stalemate than an insightful discussion. Sure, we can't fairly test other animals against ourselves but equally we cannot divine what the alternative measures might be. Regardless of the test, the findings will always have to be understood with some construct like 'intelligence' or 'culture' so we may as well attempt to find a use for these measures before reinventing the wheel. In this chapter, we'll explore a series of ways in which prominent thinkers have attempted to define humanity over the years. I'll take this space to pick them apart and evaluate each one as objectively as possible. By the end, we'll hopefully arrive at a point where 'intelligence' and 'human' are useful terms instead of utter rubbish.

Round 1: Humans are tool users

I hope by this stage in the book you're not seriously going to contemplate this. Yes, humans use tools. So do animals. If that's obvious to you, then I personally give you permission to skip this section. I promise that the other claims which follow aren't quite as ill-conceived. But for the most dedicated and/foolhardy amongst you, I will endeavour to explain.

Half a century ago, Jane Goodall returned to the UK from a trip of several years spent studying wild chimpanzees in Gombe Reserve, Tanzania. She had made a remarkable discovery: chimps use tools! Goodall had observed individuals using long grass stalks to 'fish' termites from their mounds. The stalk would be inserted into a suitable opening, agitated, then removed with the termites still clinging to it. These were then stripped off and eaten like meat from a kebab.

Prior to this, the notion of Man the Toolmaker was widely accepted in academic circles. Our ability to manipulate objects into useful devices was seen above all else as being our defining trait. No longer. Goodall's sponsor for the research, anthropologist Louis Leakey, commented on the matter: "We must now redefine man, redefine tool, or accept chimpanzees as human!" I couldn't have put it better myself, so let's start working through the options, shall we?

Are chimpanzees human? I think not. Are chimpanzees people? Now that's a different question, asking whether they're worthy of rights and recognition as separate living beings. I don't intend upon answering that here at any length, but the answer undoubtedly is 'yes.' Can we redefine 'tool'? Probably, but it's not going to get you anywhere. Hence it comes down to redefining man.

There are crows on the island of New Caledonia that fashion hooks from twigs, using these to pull juicy beetle grubs from crevices in trees and logs. In fact, the birds are so attached to their tools that they will keep and use their favourite ones repeatedly. Then there are dolphin pods observed off the coast of Australia performing a behaviour known as sponging. To hunt fish on the sea-floor, dolphins' echolocation is unreliable so they use a different technique. Ripping a sea sponge from the bottom, they hold this in front of their nose whilst they probe the sand. The terrified fish are spooked and make a run for it, thereby giving away their location. Just as humans use gloves to protect their hands when performing dangerous tasks, so these dolphins have learnt a way to shield their delicate mouthparts from damage.

The skill the crows display is in designing composite tools, working from a mental picture of their desired end-product and combining sticks and spines, or fern stalks and leaf barbs to create it. The dolphins show collaborative tool-use, often feeding in pods to improve the effectiveness of each individual's efforts and allowing them to benefit from others' actions. Likewise, orangutans have been known to create makeshift whistles from leaves, elephants fashion branches into fly-swatters, gorillas deploy makeshift bridges, and macaques floss their teeth with human hair. In the animal world, there are an awful lot of tools.

Round 2: Humans have culture

Well, at the rate we're losing languages and local customs around the world, you might not be saying this in a hundred years. Still, it's worth addressing. What is culture? Our good friend, the Oxford English Dictionary, has a dozen entries under the word. Of these, the following is most applicable: "The distinctive ideas, customs, social behaviour, products, or way of life of a particular nation, society, people, or period." It's pretty vague as you'd expect, but there are some key points to notice. Culture has to be shared by a society, hence is universal. It also has to be transmitted or transformed over successive generations. And finally, it has to be unique to a given grouping or locality. The million-dollar question is, can we really find all this in the animals?

The short answer is 'yes.' The long answer is 'yes, you blinkered fool.' Do you want the really long answer? Orcas (known as killer whales to some) are one of a handful of species that undergo menopause. Grandma orcas will stop reproducing around 40 years of age but may live well into their 90s. In contrast, males only live into their 50s and are fertile throughout their entire lifespan. Why this huge divide? Orca society is strongly matriarchal, meaning the girls rule the pod. Grandmas, therefore, carry with them decades of experience in how to find food, how to get from a to b, and other such gems. Calves learn quickly from this wealth of knowledge and have much better survival odds as a result.

These behaviours are passed down through generations, distinct to each group of orcas and applicable to the whole pod. That easily ticks the boxes to qualify as culture. Some behaviours, like intentionally beaching themselves to hunt seals, are highly specialised and require years of practice to master. This is no 'stick in a termite's nest' scenario. The biologist-author Richard Dawkins coined the word 'memes' to refer to a unit of cultural transmission. So by this logic, grandmas are the queens of memes. Who'd have thought?

Meerkats also demonstrate a culture of sorts. Adult meerkats instruct pups on hunting, bringing them scorpions on which to practice. Initially, the scorpions are dead, but as the pup improves live ones are brought with the stinger removed. Finally, the novice meerkat is given scorpions fully armed and tutored through how to eat them without getting stung. The process takes time, but with persistence and constant practice young meerkats become adept at these hunting skills. Any of the adults in the group might teach them, making this a truly selfless act.

Perhaps you don't view scorpion-hunting as culture? Ever watched Bear Grylls? And besides, it's not all that different to football. I'd watch the sport far more if they used a ball that bites. The point is, human culture is extremely diverse, and hunting, teaching, travel and cuisine are all just small components of it. Similar to how humans go above and beyond in the tools we make, the range of behaviours that characterize us is more distinctive than the behaviours themselves. Don't worry: we'll get to this soon enough. Good things come to those who wait, and the best answers I'm saving for last.

One of the most famous examples of culture in animals involves tits and milk... woah... okay... let's rewind. Blue tits are a variety of small English bird, and the milk in this instance was that delivered by milkmen in ye olden days. It just so happens that a few clever birds managed to break open the bottles' seals and so made the most of their daily delivery with a refreshing morning drink. The behaviour spread quickly from tit to tit, and soon swathes of milk bottles were being raided across the country. Often the birds would follow delivery trucks to swoop down for the freshest stuff, taking just the cream layer off the top. Nowadays, supermarket deliveries mean that this opportunity is all but extinct, and blue tits are falling in numbers.

Purists would argue that in each of the above cases, the animals' culture was simply a common behaviour that helped them to survive. What about culture for culture's sake? What about the artistry of Jackson Pollock or the compositions of John Cage, or the infamous modern-art toilet installation of Marcel Duchamp?

Bird songs display remarkable flexibility in both form and feature, with that of the lyrebird most esteemed for its beauty. Surely this virtuosic composition matches or surpasses many modern human works? There are plenty of birds that can run off arpeggios faster than the most skilled human pianists or opera singers. And I can also say without a shadow of doubt that a mouse could play John Cage's 4'33" in any given key. There's a turtle called Koopa that's made hundreds of Pollock-esque paintings, and a man named Steven Kutcher plays agent to a suite of bugs doing much the same.

If you can unravel those arty references, then you might get the sense that I'm pulling your leg. Not entirely, though, because there are parallels between aesthetics in the human and animal realms. The display grounds of bowerbirds are a textbook example- think laurel wreath gone artsy. Decorations in and around the bowers are arranged to create a clever forced perspective. Bigger items go further away so that they appear to be the same size as smaller, closer ones. The result is that the bower seems to be shorter to a visiting female, and makes the displaying male look even larger and more impressive. Forced perspective is common to many of our most impressive architectural feats as humans. The Greek Parthenon with its tapering, unevenly spaced columns is one of many examples where the effect is a perfect appearance at a distance.

Whilst bowers don't help to feed the birds, they clearly help males to find mates. Food and sex drive animal aesthetics, but that's no more than could be said for any human artist or creator. Da Vinci made his masterpieces not simply out of a burning intellect or desire for perfection but due to the more mundane necessity of pleasing patrons, gaining repute and wealth for a comfortable existence. Such status undoubtedly would have assisted him in passing on his genes had he been so inclined. In fact, the art master was homosexual- a trait also found in the animal world.

We're looking for an indisputable answer to what defines humans, so neither culture nor aesthetics can suffice. We can't deny that there are strong practical roots underlying all of our artistic endeavours. The fact that our so-called world domination allows us more free time to cultivate them is hardly sufficient grounds for calling human uniqueness. Though I do quite like the idea of humans being defined as sluggards.

Round 3: Humans have language that expresses fictions

Okay, so we're getting warmer now- and no, I don't mean Harry Potter sorts of fictions. A fiction for our purposes is anything that can't be fully understood in the physical sense. Animals love magic as much as you or I, simply because it defies their expectations. Could they tell their friends about it? That's hard to say, so let's start simple and work up to the hocus pocus of big ideas.

The Tower of Babel story in the Bible recounts how early humans came together to build a tower that would reach the heavens. Initially, these people acted with a single clear intention and all spoke a single language. God, however, saw this as a threat to his teachings and made them speak many different tongues. The project descended into chaos and the tower was left unfinished. For as long as we can remember, language has been seen as a powerful force. It's very often implicated as a driving factor in the rise of humans. Mind you, that's not the same as defining the human species. To find if it's up to that task, we need to look for language in other creatures. How much can other animals say?

Emotions are some of the simplest 'fictions' that we can express. Simplest being a relative term. Because emotions are so wishy-washy, scientists experience a rough blend of them simply trying to find order in the mess! One who's come close and achieved most clarity is a certain Robert Plutchik. He neatly set out a ring of 8 basic emotions which lead on to more complex sentiments and combine in clearly defined paths. His basic emotions were joy, trust, fear, surprise, sadness, anticipation, anger and disgust.

That's a neat place to start, isn't it? Let's take your pet dog (or cat or horse or brother) as an example because I bet you could recognize all these emotions in good old Spot. "Un-scientific" I hear you say?! Spot isn't so different from a chimpanzee, a rat, or a dolphin despite a few missing brain cells and a lack of fins. We spend more time with pets and so come to understand them better. That's all there is to it. How can we tell that animals can feel anything? We can't, but you're fine accepting it in other humans on identical evidence, so it really is a matter of all or nothing. Crying, shouting, embracing, laughing and other such displays aren't exclusive to humans. We take these as proof of emotions in our day-to-day lives. Hence we can safely say that animals feel emotions.

The term commonly applied when people discredit such claims is 'anthropomorphism.' Indeed, the Wikipedia page for this is almost (but not quite) as long as that on animal emotions. The theory goes that because we're so familiar with human feelings, we tend to attribute them to animals that might not actually show them. That's all well and good if we're talking crocodile tears, but if it's an elephant mother mourning her calf then most would say she's grief-stricken. There's a limit to how much one word can discredit, and Frans de Waal (whom we encountered last chapter) coined the term 'anthropodenial' for the opposite effect. Anthropodenial refers to describe our tendency to deny complex feelings in other lifeforms. Regrettably, the term never really caught on and my spell-checker still tries to correct it.

There are too many emotions for me to cover in depth, so let's confine ourselves to some of the most difficult. Empathy and moral virtue feature high on my list, and have become part of a new front for human exceptionalism. The theory goes that our 'circle of compassion' sets us apart: we care about more living creatures than other animals do. It's a lovely thought-no doubt about it. However, I would argue that this trait is more a byproduct of our technology (Round 4). Animals show their emotions clearly enough, communicating them through both language and posture. Koko the gorilla cared for her pet cat, and circumstantial evidence points to deep bonds between unrelated animals. Until we're able to put the question to them directly, this one's up to debate.

Chimpanzees in the wild have been observed offering consolation to victims of aggressive treatment. In other cases, chimps shown videos of emotionally-charged scenes responded strongly to these replays. Ape Hollywood might not have taken off yet, but it seems the audience is there. Macaques offered a food reward for giving companions an electric shock will happily forego this short-term gain for the benefit of the other monkey. These sorts of responses are not innate but rather learnt via interactions with others. I wouldn't trust a random human to spare me a shocking and forego a doghnut reward. Clearly the macaques know that in the long-term, kindness pays off.

On the bright side, there's a good dose of optimism to be found in nature too. Rats love being tickled, laughing above the range of human hearing. What's more, having been tickled, rats become much more optimistic. They respond to glass-half-full type tests almost always in the positive. Similar experiments have shown this sort of thinking in everything from sheep to dogs, from birds to bees. Whether animals can talk about this is tricky to assess, but we've recently taken a big step forward in breaking the code. An AI programme called DeepSqueak released in 2019 can translate rodent noises into something we can understand. This allows us to witness for the first time everything from courtship songs to domestic disputes in these charismatic critters.

Does that please you? Good, because we're moving on from emotions and on to the two most complex 'fictions': money and religion. Before I'm sent stream of angry letters, I will state that I am an atheist. Still, when I refer to religion as a fiction I mean only that it's a metaphysical belief system- a perception of reality shared by many. As for money, which we all wish was fictional, animals can turn up some surprising skills. The proof comes in the form of one of my all-time favourite studies.

You might think of Yale University in the US as a prestigious institution for the best and brightest students. It's a leading member of the Ivy League, and ranks 8th in the World University Rankings 2021. So what if I told you that two of their students taught capuchin monkeys to gamble? "Science" manages to excuse almost anything!

For the test, a dozen monkeys were instructed on the use of a fake metal currency that they could trade in for food rewards. This first stage all went pretty much to plan, and the monkeys surpassed all expectations. Their budgeting abilities proved to be on par with humans, adapting their spending habits to changing prices as economic models dictate. One of the experimenters went so far as to say that the data they generated was "statistically indistinguishable from most stock-market investors."

As you might expect, the clever creatures then progressed to bartering, stealing and gambling in the natural succession of things. One sneaky individual somehow got its hands on an entire tray's worth of coins. Chaos ensued in the enclosure, and it was amidst this general kerfuffle that one capuchins was seen paying a lady monkey for sex. That gives her a claim for being the first known non-human prostitute, as she promptly traded her earnings in for a grape. This may sound disgusting, but it's also very clever. To come up with such an arrangement requires not just a firm understanding of the currency, but the ability to negotiate for a delayed reward using money. In short, language was used.

For religion, it's best to start with a definition. The "non-anthropocentric, non-anthropomorphic, non-theistic, and non-logocentric trans-species prototype definition" used by researchers probably falls under the category of being little too wordy for our purposes. Back to the Oxford Dictionary, and here we find the bland description: "Belief in or acknowledgement of some superhuman power or powers which is typically manifested in obedience, reverence, and worship." Good enough. So do animals show belief in gods or magic? Probably.

Elephants have been known to display a sort of "moon worship," performing ritual bathing when the moon is full and waving branches at it as it waxes. Chimpanzees burst into dance when heavy rains fall, and do likewise if they come across a waterfall. These actions in themselves aren't full proof of religion per se, but they're very close to it. Elephants can talk to each other 10 kilometres apart by speaking in low rumbles they pick up through vibrations in their feet. Who knows what exactly they're discussing?

Round 4: Humans have flexible cooperation over global scales

Well well well. This one really is a whole new ball game. I'm sick as a parrot from hearing poor arguments for humans being different, but with a statement like this, our discussions can go somewhere. There's plenty to cover, so let's take this one step at a time. I've had an albatross around my neck trying to shine a light on the simple previous points, so goodness knows where we'll get with this one. To set things in motion, let's ask whether animals can a.) cooperate, and b.) cooperate flexibly?

By now, you've probably worked out I won't be answering such a question directly. So here's a story to drive the message home. Akeakamai and Pheonix were two dolphins involved in some early language experiments. Both were taught to recognize human words, but one was taught via hand gestures and the other via sounds from an underwater speaker. Once they had mastered this, each was given the instruction to perform a trick with its partner. Sure enough, the two got together and performed their motions synchronously time and again: diving backwards or splashing their tails or any other random trick as the command was repeated. These actions were untrained, and therefore spontaneous. That means they could communicate their intentions to each other, and also display creativity by mixing up the displays. This is flexible collaboration in a nutshell.

As for scale, in humans the largest form of cooperation is seen in countries forming international networks. Hence the stipulation that cooperation occurs globally. Ants are famous for their social organization- the Argentine ant perhaps more so than any other. How do the two compare? Most ants are content to live in single colonies, like cities. Argentine ants, meanwhile, are empire builders. They're kind of like Portugal in the age of exploration, founding colonies across the globe. The difference is that they're much, much more successful than any single human empire.

Originating in South America (there's a hint in the name), Argentine ants have spread to every continent except Antarctica. Each world region harbours an Argentine ant 'supercolony,' the members of which are all genetically related. The largest of these stretches for 6000km along the Mediterranean coast, encompassing multiple countries, millions of nests and billions upon billions of workers. Maintaining friendly relations in human politics is a challenge, but ants from different supercolonies will greet each other amiably. That's an enviable level of social cohesion that humans should aspire to.

There are about 100,000 ants for every human on this planet. These tiny insects weighed more than all the humans combined until about the 18th century. So if a single one of the 20,000 ant species can so dramatically challenge our views, it begs the question of what their close relatives might have to teach us too. More to come in chapter 6 on this...

The difference between humans and ants, you might say, is twofold. Excepting the fact that humans are thousands of times larger and lack an extra four legs, we have politics and we can actually talk to each other internationally. We can debate future paths and select the most appropriate, with elected representatives standing for the will of the whole. Ants can't match that. Plus there are those that would argue ants can't talk at all!

They can. They do. So those people would be wrong. Ants talk by rubbing special ridges on their bodies a bit like a xylophone glissando. Chatting across continents is another matter, but just because no-one has let ants onto Zoom isn't their fault. Given what's been shown about their peaceable interactions with foreigners, they certainly could talk across countries if provided with a suitable video conferencing solution. Technology becomes the decider in points like this, and I'll be addressing that elephant in the lobby very shortly.

George Orwell's *Animal Farm* aside, there's plenty of politics to examine in the animal world too. Going back to ants, it has been shown that the division of labour system used by colonies follows the same governing processes that bring about factionism in modern politics. 'Social influence' describes how individuals become more like those they interact with, and 'interaction bias' speaks to how we tend to interact more with individuals similar to ourselves. Combined, these forces mean that society naturally segments into polarized political groups: Liberal vs. Labor in Australia, Republicans vs. Democrats in the

US, Conservatives vs. Labour in the UK. Ants don't need lengthy rhetoric and debates, but they do need an effective means of task delegation. Instead of fixing workers into a particular task, they achieve this simply by keeping busy and following their natural tendencies to enjoy certain roles.

Now, there have been plenty of political agendas expressed in recent history, not all of them hugely successful. Dare I say if we attempted an almost completely unregulated society such as these ants prosper in, it would implode within days. Yet these tiny creatures excel at it, and in doing so have become some of the most successful lifeforms on this planet. Sometimes I want to give our political leaders a magnifying glass just so they can learn from these amazing critters. Think about the elegant efficiency of this system of theirs and then tell me we're the better species thanks to blasted politics!

But on to the devil of technology. Ever wondered what it would be like if two chimps connected via FaceTime? Well today's your lucky day because I've got some wonderful news for you: somebody's tried it to save you the trouble! This beautifully mad idea began a little under a decade ago with a joke: "a singer, an MIT professor, a telecoms whiz and a dolphin psychologist walked into a bar." At least that's how I imagine it, and I'll leave you to come up with the punch line. This strange group of people came together over an idea to build what they called an 'Interspecies Internet.' In simple terms, this would have meant apes and dolphins and elephants talking to each other via the world wide web- and to people too once we could translate what they were saying. The idea has been slow to gain traction, and the technology is lacking, but a few small trials have already taken place.

Which brings me back to chimps on the phone. Working

with a rescue centre in the UK, the team managed to hook up various chimp pairs on FaceTime calls. What came out of it was hilariously predictable in hindsight. Given this new tech toy, the chimps had the time of their lives playing games, trying to pass a banana through the screen, and generally making a fool of themselves. All in all, the experiment was a resounding success and, bananas aside, the cheeky chimpanzees had little difficulty grasping the concept of the magic portal they seemed to possess.

Is this the same as having invented video conferencing? No, of course not. But even so, it continues chipping away at this great mountain of our self-esteem to know that other animals can use our iPhones. And if it's technology that sets us apart from other animals, does that make your grandparents less human because they lived before the internet was around? Or think back to the 18th century, before the telegraph- where do we draw the line for when humanity starts? Certainly, before the advent of farming there was precious little in the way of technology that set us apart.

Using the same logic, you could argue that our humanity is increasing in line with Moore's law. Do you really believe that to be the case? Project forward two-hundred years, when we've colonized Mars, engineered telepathy and toyed with immortality, hopefully not destroying our planet in the process. Now, once we have gained all this power- perhaps even inflated our intelligence with AI implants- now are we still human? Where exactly does "super-human" start and when do we reach peak human? The problem with technology is that it's not as constant as we wish humanity to be, so sadly this is not and cannot be the answer. One last round to go...

Round 5: Humans are opportunistic ecosystem engineers

Okay, this one I've never heard except from my own mouth. I guess that's where invisible friends come in: as someone to reaffirm your brilliance. This definition I will argue in favour of, in part because I made it, but mostly because I sincerely believe it to be true. Whilst many of the other rounds focused on the present state of our species, this seeks to define the specific traits that have allowed for our success. I've used biological terms on purpose to reaffirm that we are still animals, but if you'd prefer the plain English version, then read the above as 'Humans are thrifty troublemakers.' Either way works.

Now, the Australian Indigenous peoples possess the oldest surviving culture to date, so we'll start there. 50,000 years ago, shortly after the first humans left Africa, their predecessors reached the Australian mainland by island-hopping from Asia. When these ancient explorers first set foot on the continent, it was inhabited by prehistoric megafauna. There were wombats the size of large SUVs hunted by marsupial lions and 7m long predatory lizards. Not to mention the infamous Demon Duck of Doom standing 3m tall. What caused these creatures to go extinct is a matter of some contention, but evidence suggests certain Aboriginal practices may be largely to blame.

Australia is a land designed to burn- now more than ever. When Indigenous Australians first arrived, they were quick to realize this and use it to their advantage: systematically burning patchwork areas of bush across the country to drive our prey for hunting. Afterwards, as the fresh growth came through, many more animals would gather there, increasing the richness of their pickings. Because of this brutally effective technique, the slow-moving megafauna were often on the menu. Dangerous carnivores like the lions and lizards were hunted in self-defence and to reduce competition, whilst the slow reproduction rates of all the megafauna stunted their recovery. Where rich rainforests had once covered much of the continent, drier scrublands now took their place. The bushland evolved to burn more readily and recover faster.

To paraphrase, these early peoples set about cleverly reconstructing the landscape into one more amenable to their existence. The complete removal of megafauna wasn't intentional, but the ecological changes they created largely were, and the result was that they created habitat fragments which they could use to their advantage. Remember that an ecosystem engineer is any species that substantially alters its environment, and an opportunist is one that rapidly claims the new niches such alterations create. This in essence is what we humans have done throughout history across every continent. Neither trait is unique in itself, but the combination adds potency.

Another example that might not occur to you in the modern context is production lines. The computer that I'm typing out this manuscript on has components from around 40 countries. It has a screen from South Korea, a drive from Thailand, raw components like steel from Russia and copper from Chile, glass from Korea and lithium from Zimbabwe to name but a few. I'm a pretty grudging commentator where technology is concerned, but I've got to admit that's one hell of a feat. Just as ants specialize for roles within colonies, we have spent the past two centuries setting up convoluted supply chains for the products we depend on. We alter environments by extracting raw resources, and use these materials to better feed our 'world domination.' The former is acting as an ecosystem engineer, the latter as an opportunist, but the two of them act hand in hand.

Painted this way, it might seem that development is at odds with sustainability. I've never been much of an artist, but let me round out the chapter by painting you a slightly different scene. Being 'human' has been politicized far beyond our core traits and values. We've doubled our global life expectancy since 1900, but people are still dying so we feel it's insensitive to celebrate. We've gone from almost none of Earth's land area protected to nigh on 15%, but biodiversity loss is rising so still we feel we can't congratulate ourselves on this victory. There's a term coined by the late author Hans Rosling, and that is 'Factfulness.' It refers to the ability to recognize global issues but take faith in prior successes, merging what we can learn from the past and present with an optimistic future outlook. So open your mind ever so slightly, strive for factfulness and I'll give you a sneak preview of what Human Nature could mean.

In the 1960s, US space agency NASA funded a pioneering attempt to teach dolphins to speak English. It had been observed that the animals could imitate humans with considerable accuracy, and so a room was flooded, a dolphin called Peter introduced to his new home, and a naturalist named Margaret Lovatt took up residency on a platform suspended above. After many months of trying, Margaret had taught Peter a handful of English words but conversations were a long way off. The project faltered and funding was withdrawn, leaving Peter's language lessons unfinished.

The stated intention of these efforts was to school dolphins in the English language to such an extent that they could take up a seat at the United Nations on behalf of marine mammals. Would this be possible today? Yes, no, maybe and soon. Yes, we have got further in conversing with dolphins in the half-century since Peter and Margaret first spoke. No, we haven't yet been able to create a dolphin primer in human politics. Maybe if they were better-versed, dolphins might yet take up a UN seat, and soon (in a few decades' time) our technology might be able to act as an interpreter. 'Humanners' is the heart of Human Nature, and refers to the basic courtesy of treating animals as we do human beings. With this in mind, we must consider the above as a possibility.

CHAT stands for Cetacean Hearing And Telemetry. The box to which it refers resembles a futuristic iron crossed with a toaster. When strapped to a swimmer, however, it allows for dolphin language to be partially decoded. Dolphins swimming with researchers grow accustomed to human words and equate these with their own language of clicks and whistles. The results are recorded and slowly a sonic dictionary is built, allowing people to play back recorded dolphin words and start to converse in dolphinese.

In 1977, when the voyager golden record was launched, whale song featured amongst the material on the disk. This joins other entries in a condensed version of human culture making its way out towards the stars. If we're prepared to initiate respectful dialogue with aliens and to use whale song as a common language, then we should be more than comfortable listening to what Earth's other lifeforms have to say. An animal representative at the UN would be quite a feat, but in the meantime humans must act as voices for the 'voiceless.'

Just as non-European peoples were abused in the age of exploration, animals have had some pretty rough treatment over the past few centuries. By and large, we've now accepted that all races deserve human rights. We've begun to respect the wisdom of Indigenous peoples and we've embraced them in modern society. Custodianship of traditional lands is being recognized, and formal apologies have been issued for historical wrongs. Human Nature dictates that the same should be possible for all animals, with a reconciliation of harmful practices and recognisal of Humanners. It's not beyond imagining.

Human ecosystems are part of our modern world. Most of the Earth's land surface is affected by our activities, so as well as protecting the wild places that remain, we need to shape our own settlements, farms, towns and cities to allow other animals to thrive. Human Nature means showing respect, appreciation and compassion towards other lifeforms. That includes making space for them alongside us. Pigeons and ants and rats are doing fine as is. The test comes in whether I can see deer wandering past Buckingham Palace, cockatoos perched on the Sydney Opera House, or a snail climbing up the Eiffel Tower. Any of these will show that Human Nature has arrived.

So I'd like to see critters welcomed into our cities, remedying of animal exploitation, and refinement of sustainable development to consider animal wellfare. Mahatma Gandhi once said that "The greatness of a nation and its moral progress can be judged by the way its animals are treated." By that measure we've got a long way to go, but things are certainly improving.

A little while ago in a galaxy called the milky way, and on the planet we fondly call home, there was a monkey named Naruto. This was no ordinary monkey, mind you. Naruto was a monkey with a beautiful grin and a particular love of one modern fad: the selfie. Naruto was out minding his own business when along came a photographer snapping off pictures of him and his fellow macaques. Not wanting to miss out on the fun, Naruto promptly grabbed an unused spare camera the photographer had set aside and snapped off a few cheeky pics. The results were top-class by any standards.

When the photographer got home from his travels, he added Naruto's selfies to his own portfolio, selling prints of the monkey's priceless grin at a premium. Except that the animal rights organization PETA caught wind of this tale and took the photographer to court- suing on Naruto's behalf for copyright infringement. They lost the case, but in return, the photographer agreed to pass on a portion of all profits to macaque conservation to help out monkeys like Naruto in the wild. So the simian celebrity was rewarded for his troubles even whilst animals took another blow in court rulings. Who knows- perhaps someday we'll recognize the intellectual property of animals?

Before Naruto's case hits the courts, when Disney was producing their first *Lion King* film, another similar drama took place. To gain familiarity with the animals, a team of animators spent several days studying and sketching real-life hyenas. American hyenas, that is- some captive specimens held at the University of California's Field Station for Behavioural Research. Several of the hyena researchers there accompanied the artists with a very clear goal in mind. They were fed up with how hyenas were seen as cruel, stupid beasts and begged the people from Disney to show hyenas in a positive light.

If you haven't seen the film, then take my word for it- hyenas couldn't be made to look any more cruel or stupid. And the scientists were furious! One of them later wrote in an article for Africa Geographic that one way people can help hyena conservation is to boycott watching the Lion King. Another researcher went so far as to sue Disney for defamation of character, acting on behalf of the hyenas! As you can imagine, the case didn't get very far. After all, Disney has some of the best lawyers in the world to defend them. But, as the saying goes, it's the thought that counts.

Other cases have seen animals granted certain rights. In 2015 an orangutan named Sandra was deemed a legal person and released from captivity to a new home in a sanctuary. Rivers have received similar treatment in South America and New Zealand, with their rights legally recognized and represented by local Indigenous people. Humanners shows this simply to be good sense, and I hope that Human Nature means more recognition is given to animals moving forward. The momentum behind this transformation is mounting, and once it is complete there should be little need to bring nature's interests to court for a ruling. People are coming to value and understand the human-ness of animals, and whilst it's agreed intellectually that humans are animals, the inverse is more meaningful. Animals are individuals, are people, and should be treated as such. So, with that said, it's time to take out the magnifying glass once more and focus down on some of the wonderful creatures that share this Earth with us.

6

No-one's Ark

My body is worth $1,500. If you were to blast me with some hyper-advanced atomizer, reduce me down to small lumps of different elements and sell them on, I'd be pretty peeved. But that's how much you might expect to make. If you prefer, you could make 3 million matches out of me, or haggle various organs of mine for large sums on the black market. Never has the phrase "don't try this at home" been more pertinent.

The point of this hypothetical is to say that the beauty is in the detail. On the chemical level, there's very little that differentiates me from a cow or a coprolite (science-speak for fossilized dinosaur poop). That's why we flatter ourselves by looking at the big picture. Evolution is a simple enough concept: things change, and often good changes stick. However, given billions of years in which to work its magic, evolution produced the panoply of lifeforms that grace this Earth today.

Animals make us human, and when we look at nature we see ourselves. So in a book about humanity's relationship with nature, it would make little sense not to dive into the beautiful detailed work of biodiversity. As we pause from the wider narrative of Human Nature, I've chosen ten creatures to remind us of who we are. Many of them are threatened with extinction, and I've labelled them all with their conservation status. More importantly though, all of these animals have a story to tell about humanity's connection with the natural world. We'll follow each one in turn as they join us in our ark, and we'll examine what each of them has to teach us about ourselves. So it's anchors aweigh as we set under full sail, and may the trade winds blow us steady!

1.) Wandering Albatross (Vulnerable)

Diomedes was one of the great Greek heroes, described in Homer's *Iliad* and deified thereafter with a suite of myths to his name. For instance, he once wounded the god of war (Ares) with a spear- that's no easy feat! Legend has it that, returning home in triumph from battle, his ship was blown off course in an almighty storm and driven onto the coast of Southern Italy. Stranded, beaten, and battered, he languished there for the remainder of his life, his companions turning into great white seabirds on their passing. Diomedeidae is the scientific name for the albatross family in homage to this legendary tale, and the birds themselves have long been the subject of myth and superstition.

One example is found in Samuel Taylor Coleridge's <u>The Rime of the Ancient Mariner</u>- amongst the best-known works of English poetry. In it, the narrator kills an albatross with his crossbow. Fearful of the ill portent this act represents, his fellow sailors make him wear the bird's carcass around his neck. According to legend, seabirds like albatrosses were the embodiment of lost sailors' souls. Hence the sight of one meant good fortune, and shooting an albatross would incur the wrath of the sea. The same idea is found in Herman Melville's <u>Moby Dick</u>, with an entire chapter bearing the bird's name.

Before the age of exploration came the age of speculation. Few had the means to journey abroad, and so hearsay predominated in popular culture. This was the earliest form of natural history, long before the scientific discipline had arisen. 'Knowledge' was a mixture of patchy observation and campfire tales, examined only where it impacted on daily life. Precious few animals bore the honour of description in these early days, and those that did were those of special significance.

What makes a species noteworthy? Modern science shows that it must either a.) look like a human, or b.) act like a human. Or both. TV broadcasters and conservation charities recognize this, hence the repeat casting of a small number of species in documentaries and donation appeals. We came across these flagship species in chapter 3, so flick back there if you need a quick refresher. 'Cinderella species' are those which are similarly loveable but close to extinction and ignored by the public: the pygmy raccoon for example, or the Pennant's red colobus monkey. But these are the exception, not the rule, and on the whole, alien-looking cockroaches get stepped on whilst chimpanzees are liked by people and get treated okay.

As for albatrosses, they don't look much like humans, but their actions more than make up for what they lack in appearance. One satellite-tagged individual was recorded flying the distance between London and New York in a mere 12 days at sea, and others have been known to fly three times as far gathering food for their young before returning to the nest. We love this sense of freedom and the grace of their flight across the sea- add to that their parental devotion, with couples staying together for over 50 years, and we cannot help but admire these beautiful creatures. They also have the largest wingspan of any bird, greater than the height of a basketball hoop. If you want people to notice you, being big certainly helps!

Unfortunately, it hasn't all been smooth sailing for albatrosses. When sealing and whaling vessels arrived on the sub-Antarctic island of South Georgia in the 19th century, they brought with them brown rats. The damage was akin to killing two birds with one stone: whilst whale and seal populations were decimated, rats invaded the island and set about feasting on the young of the birds that nested there. Millions of birds representing 30 species are known to breed on the 170km-long stretch of island. The conditions are too harsh for trees to grow in, so albatrosses and other seabirds nest on the ground or in burrows. That means their eggs and chicks are easy prey for hungry rats.

That's the bad news. The good news is: they're gone. The rats, not the albatrosses! In 2018, the South Georgia Heritage Trust gave the long-awaited news that the island was officially rat-free. With a ten-year eradication (or "e-rat-ication?") programme costing $13 million, hundreds of tonnes of poison were air-dropped across the non-frozen patches of ground, systematically removing the vermin. It's a sad waste of life for the rats that were killed, but it was our fault entirely for bringing the creatures onto the island and thus our responsibility to negate the damage caused. Overall, countless more animals benefitted, with the vulnerable bird populations rebounding in the few years since.

This scheme was eight times larger than any eradication programme previously attempted. That's quite an achievement, and proof that if we put our minds to it, we can combat large-scale problems. Longline fishing hooks kill one hundred thousand albatrosses annually, but through our human ingenuity, we've arrived at a solution. A brand called Hookpod has patented a system for a casing that releases fishing hooks only once they reach a depth of 20m. This is beyond the reach of almost all seabirds, thereby drastically reducing fatalities from these creatures getting snared by the lines. Even HRH the Prince of Wales has spoken out for albatrosses, developing a passion for them during his navy years.

As always, nature is one step ahead of the game in engineering for efficiency. Humans have invented Hookpods, but albatrosses are continuing to teach us about energy-efficient flight. Now, I'm going to take a bit of a gamble here and say that you can't fly unaided across the Atlantic Ocean. If you can, I'll happily buy you a cup of coffee. The reason that albatrosses manage this is that they are infinitely better aviators than ourselves. We made it to the moon: big deal. What makes albatrosses special is that they can do it all without burning the 2500 tonnes of rocket fuel used to get each Apollo mission into space. Nature is shaped by evolution to conserve energy for survival- we have a lot to learn from this.

It starts with the wings (duh). Having a large wingspan helps, and going one better than that, albatrosses have a tendon that allows them to lock their wings in position during flight. A bit like if you were melded to a hang glider.

The other trick that albatrosses use is called dynamic soaring. Essentially, a bird will read the local air currents and plot a course, using headwinds to gain lift and then gliding slowly down, tacking to and fro, and angling themselves to maximize their distance travelled. They glide at speeds of 40 or 50 kilometres per hour and barely have to move a muscle in doing so. There are drones in the works that will employ a similar principle, but the complexity of the skill means that these fliers are a few years off yet. Even the wing colouration (black on one side, white on the other) plays an important role in the albatrosses' flight which we're just beginning to explore.

I've included the wandering albatrosses on my ark as a reminder that there is still much to be learnt by looking through another creature's eyes. Their relationship with humans is long and varied, but recent measures mean that these magnificent birds have much more hope for the future.

2.) Matabele Ant (Data Deficient)

The Matabele were a fierce Zulu tribe that fought against colonial conquest in a series of brief but brutal wars. Their name comes from the tall traditional cowhide shields they used to defend themselves in combat. Sadly, cowhide does little to protect against a shot from a bayonet, and the tribe was heavily outgunned. However, they were respected and feared at times by these invading powers for the courage and cunning they displayed in their warfare. What better name could there be for an ant species that excels at military strategy and prowess?

A raid begins with one of the ant scouts leaving the nest, spending an hour or more traversing the human equivalent of several dozen miles on foot in search of a target. The prey they seek are termite nests, and if one is found, the ant will make a beeline back for the colony, leaving a trail of pheromones (scented chemicals) behind it. Upon arrival, it takes a mere 60 seconds for the army to assemble, and hundreds upon hundreds of ants stream out along the scout's marked path. They reach the termite mound and breach its walls, overpowering any defence that is mounted. The goal is to capture as many of the termites' larvae as possible, dragging them away back to the ants' nest as food. Often, the raid is over in a matter of hours, but there are always casualties. What of the wounded?

Large-scale medical treatment had never been observed in the animal kingdom until recently. Yet fresh observations show that during and after the heat of the battle, injured Matabele ants are rescued by their comrades, carried back to the nest and given emergency first aid. Nurse ants treat injured soldier ants' wounds by licking them thoroughly for several minutes to prevent fatal infection from setting in. The simple treatment is remarkably effective, with 10% of those cared for passing away as opposed to 80% if left untreated. That's a strong record by any measure, and it gets even better.

Whereas in humans, emergency doctors have to select the patients most deserving of their attention, these ants are selfless enough to direct their carers. If they are mortally wounded, they will kick and struggle and generally refuse to cooperate with any ant who attempts to care for them. If their injuries are curable, however, they actively play up their wounds: limping when other ants walk by, and emitting a special distress pheromone to attract attention. Once treated, many of the ants go straight back to fighting- a display of courage few humans could muster.

In spite of their size, ants achieve impressive feats through the uncanny cohesion of thousands of workers. Simple rules followed by many smaller parts sum into larger complex actions known as distributed processes. Individual neurons firing in your brain are a good example, allowing you to do amazing things like reading this book. When it comes to ants, science has started viewing each colony as a single 'superorganism,' due to the remarkable efficiency with which the ants cooperate. That's one heck of a claim!

Considered this way, ant superorganisms are some of the most powerful ecosystem engineers we know, shifting incredible amounts of plant matter and soil to create their homes. We can pick up some top-notch advice from them as to how our own technology might work: swarm robots that mimic ants to complete large-scale tasks; self-healing materials that instantly repair any damage; even traffic control protocols designed for congested areas.

Leaf-cutter ants are a fine example of ant-ventiveness and ant-genuity. Did you know that they invented farming 50 million years before us? I expect not. Made famous by a score of documentaries, these ants take plant material in little confetti chunks back to their nest where they chew it into a pulp and pile it up in mounds. Add to that ant faeces as fertilizer, and what do you get? Delicious domesticated fungi found nowhere else in the world except for these ant nests, and also conveniently the only thing the ants can digest.

Other ant species take the dairy farmer approach: guarding herds of aphids (tiny bug critters) that drink sugar-rich sap from plants. High in sugar but low in nutrients, the sap must be consumed in large quantities for the aphids to sustain themselves. And the ant farmers are more than happy to suck up the excess excreted out of the aphids' rear ends. So sought after is this sweet secretion (known as 'honeydew') that the ants will regularly 'milk' the aphids for it. The farmer ants not only protect their herds but destroy the eggs of potential predators like ladybirds, and will carry the aphids between plants when one runs dry. Over winter, they store the precious aphid eggs in their ant nests and return the hatchlings to their plants come spring. Some will go so far as to clip the wings of any aphid that should grow them, preventing their precious livestock from escaping.

There are times, of course, when ants get the butt-end of a partnership: caterpillars and stick insect eggs, for example, that con the colony into caring for them. And there's the altogether-nastier zombie fungus that enslaves ants' bodies via mind-control whilst eating away at them from the inside. But overall, ants are the masters of working with nature to put themselves on top, and the relationship between the acacia ant and bullhorn acacia tree proves it.

Bullhorn acacia trees do a great service for ants, providing shelter in their hollowed-out thorns, sweet nectar from specially evolved nodules, and nutritious brown pods that ant larvae love. In return, the ants take measures to ensure the plant not only survives but thrives. They attack any animals attempting to feed on its leaves, and go so far as to sever and poison the stems of surrounding plants competing for sunlight. It's a win-win scenario for the ant and its host plant that cements their place in the ecosystem.

The Matabele ant wins its spot on my ark to show how remarkable complexity can be found in the smallest of lifeforms. Humans aside, ants control the Earth and these incredible superorganisms are only just being understood.

3.) Northern White Rhino (Critically Endangered)

Described on his Tinder profile as "the most eligible bachelor in the world," Sudan the Northern White Rhino passed away in

2018. Suffering from crippling illness, old-age, and infection, his carers were forced to euthanise him in a tragic end to years of valiant protection. His previous 24hr guard was dismissed, and media headlines around the world ran stories about the passing of a species. Sudan was the last male Northern White Rhino. His daughter and granddaughter are all that's left now, and neither of them has been able to carry a pregnancy to term.

It doesn't get more depressing than that. In the last years of Sudan's life, the Ol Pejeta conservancy that housed him shifted mountains to try to prevent the unthinkable. The Tinder profile set up for Sudan was part of a $10 million fundraising campaign. From this, a cutting-edge cell culture project was launched that could potentially save his kind. Sperm from 12 other males has been preserved in frozen form, and if this can be paired with one of the females' eggs then a rhino embryo could be born. A female from a related rhino species could easily act as a surrogate mother, raising the infant as her own. It may sound far-fetched, but that's because there simply aren't any options left. As Sudan's profile said: "I don't mean to be too forward, but the fate of my species literally depends on me. I perform well under pressure... 6ft (183cm) tall and 5,000lb (2,268kg) if it matters."

In the years since Sudan's passing, little progress has been made. Despite bringing together a team of top scientists from 5 continents, no rhino embryos have resulted nor is a pregnancy imminent. Sudan's carers say of him, he "stole the heart of many with his dignity and strength." If science succeeds in raising the dead, it will be a true miracle and testament to Sudan's legacy. As it stands, though, the Northern white rhino looks to go the way of the Western black rhinoceros, which was declared extinct in 2011.

Rhinos have been the face of conservation for decades, so it's hardly surprising that they're battling extinction. Whether the Northern white rhino is a species in and of itself, or whether it ought to be grouped with its Southern counterpart is a matter of some debate. What is indisputable is that these animals used to roam across East and Central Africa for millennia, and now none remain there. The white rhino of Southern Africa fares better with some 20,000 to its name, but shockingly this species too was once on a knife edge.

As the 19th century drew to a close, the Southern white rhino was thought to be extinct. With hunting and habitat loss forming a fatal duo across the entirety of its range, few were surprised at the loss. As it happened, though, the species was not quite extinct. Twenty surviving rhinos were found in a remote patch of South Africa, and the area was promptly set aside as a national park. From its establishment in 1895, the Hluhluwe-iMofolozi Park- the first nature reserve in Africa- stood as a beacon of hope for the Southern white rhinoceros. Slowly, rhino populations recovered under the protection they were afforded, provided with a pristine environment in which to thrive.

By the time 1950s came around, the stage was set for expansion. Operation Rhino was formed by the national park authorities, taking rhinos from this now-stable population and introducing them to safe zones across the country and abroad. For the few fortunate souls who have never wrestled a rhino, they aren't easy to move. Starting with a blank slate, new methods were devised to transport such bulky beasts over long distances. Tranquiliser darts were often shot from a helicopter, big crates truck-loaded in, a blindfold tied around each animal's head, and lots of strong men with ropes used to do the heavy lifting. Though the rhino population is still less than 10% of what it once was, the improvement from 1900 to the present day is thousandfold. The only blemish to their recovery has come in the form of poaching.

If you knew an animal was worth half a million dollars, would you kill it? Perhaps not. What if your family was starving, living in a mud-brick house, unable to afford a torch to see the snakes on the ground at night? The motivation is obvious, greed infectious, and rhinos make temptingly large targets for any man with a gun. Several 'rhino wars' have been announced when poaching numbers soared, meaning heightened conflict between authorities and poachers.

The first of these was in the late 20th century, when the oil boom in the Middle East saw a spike in demand for ivory. Wealthy businessmen in Yemen sought out the material as a status symbol. Larger demand meant higher prices, and so more people were tempted to try their hand at the black market trade. The crisis eventually abated with better rhino protection and civil war in the Middle East paring down demand for the product. But over a decade on, the second rhino war hit Africa when an ivory craze popped up in Vietnam and created a whole new flourishing market. As you'll know from the quiz you took in the Preface, this is beginning to subside, but the threat of poaching is never far away.

On a positive note, the plight of the rhino has sparked some truly innovative solutions to be devised. Fake horns have been made out of horse hair, with the intention of disrupting the market and reducing demand for the real thing. Keratin, the material that makes up rhino horns, is also found in horse hair, birds' beaks and human fingernails, so the idea isn't as far-fetched as it sounds.

One South African citizen took this to another level when he sent all his toenail clippings to the local Chinese embassy in regular deliveries, encouraging others to do the same. Although the idea (understandably) never took off, it did work as an excellent publicity stunt and undoubtedly left the embassy staff extremely confused. Truth be told, other anti-poaching measures are much sadder: dehorning rhinos or injecting their horns with poison to make usage in traditional medicine unsafe.

Whenever a species is revived from a few individuals, the genetic diversity of the population is severely compromised. Northern Elephant Seals now number 239,000 all of which have descended from 20 original survivors of large-scale hunting. In New Zealand, a single female black robin can trace her lineage through all of the 250 black robins alive today. Rhinos likewise are at a severe disadvantage from their limited gene pool. This makes them less able to adapt to changing environments, and more likely to produce offspring with crippling defects. Prevention is better than cure for declining animal populations. The challenge comes in being cognizant of all threats an animal faces and redrawing our goals as creating healthy ecosystems- not warding off extinction.

I add the white rhino to my ark as a reminder that conservation is a fight against the less scrupulous side of ourselves, and that hard-won victories can be the most rewarding. The Javan and Sumatran rhino species have only 100 animals between them, so we must learn from Sudan never to repeat our past mistakes.

4.) Dumbo Octopus (Data Deficient)

3000 metres under the sea, everything is dark- so perhaps it's not surprising few people have seen the dumbo octopus. Those who have describe the webbing between its tentacles giving it an umbrella shape, and of course the fins on either side of its head paying homage to a certain flying elephant. The name 'dumbo octopus' may refer to any one of 15 species, but all of them share these common traits alongside two small, raisin-like eyes and an off-white colour.

When the IUCN hasn't classified how threatened a species is, it can mean one of three things: a.) it was recently discovered, b.) it's smaller than a matchbox, or c.) it lives somewhere people don't tend to visit. The Matabele ant from number 2 falls under the second category, whilst the dumbo octopus scores double with a.) and c.) criteria. The pitch-black depths of the ocean aren't the friendliest of places to visit, and as a result, the dumbo octopus is a relatively recent find. We don't know how many of them there are, and can only assume that they are found across the globe. Octopuses have always had an aura of mystery to them, but none more so than these.

If I told you to draw a species with three hearts pumping blue blood, with 8 appendages each controlled by their own brain, no skeleton to speak of and the ability to match its appearance perfectly to its surroundings, chances are you'd hand me a neat sketch of an alien. And with good reason: the idea of such a creature is one that we with our limited human experience have immense difficulty comprehending. Yet octopuses have graced this planet with their presence for the past 300 million years, puzzling great thinkers since science began. Aristotle dismissed them out of hand as unintelligent, but if their captive escapades are anything to go by, we shouldn't be so quick to write them off.

In a California aquarium, one octopus deftly disassembled a water recycling valve in her tank overnight. 750 litres of water were dumped over the newly installed flooring beneath, leaving the cleaning staff with a nasty shock waiting for them the following morning. Another octopus in the UK learnt to break out of its tank at night and snack on fish from a nearby tank before returning innocuously in time for the keeper's morning arrival. What a devious creature! Finally, topping it off, an octopus can claim to have bested the human world record for solving a Rubik's cube. The time it took to solve the cube was a mere 5.24 seconds- a fraction of a second better than the record at the time. Having eight limbs might be construed as advantageous, but the feat is impressive none-the-less!

Camouflage is the ability octopuses are most famous for. Going beyond simple colours and patterns, many species are able to change the texture of their skin to match with their surroundings. Tiny moveable bumps on the surface are altered as necessary for the desired effect, allowing a pebbly sea-floor, rough kelp forest or sandy bottom all to be replicated with ease. The mimic octopus takes it one step further by contorting its flexible body into the shape of other sea creatures it chooses to masquerade as. All of this is firmly in the scientific spotlight as we decipher how these effects are achieved, but the fashion industry might see some classy new additions if we can master the underlying mechanisms.

Aside from their peculiar anatomy and unlikely intelligence, octopus behaviour is quite the enigma in itself. Previously thought to be strictly solitary animals, octopuses have recently been observed living together in sizeable communities and even larger nursery groups. Oclantis and Octopolis might leave something to be desired on the naming front, but these two locations South of Sydney on Australia's East coast represent the first evidence for octopus cities. The (also poorly named) gloomy octopus can be seen there in the dozens, sharing lodgings and getting along on quite good terms with their flatmates. One of their apartment blocks has been built around a scrap of metal that washed in, whilst the other seemingly arose from scratch using clam and scallop shells for walls. Octopuses may not be quite so solitary after all.

On the opposite side of the world, off the Californian coast-line, another recent sighting showed one thousand female octopuses gathered in a single spot to raise their young. Apparently, the joint benefits of an optimal location and protection in numbers persuaded the mothers to set aside their differ-ences, leading to this rare spectacle. That's incredible enough without the devotion of octopus mothers to consider: starving themselves for up to two years whilst incubating their eggs, protecting their precious offspring and wafting them with oxygenated water to speed along their development. When the eggs hatch, the mother's body shuts down, its cells dying en masse staring at the eyes and moving outwards to her other tissues and organs. The young octopuses take their first meal from their mother's corpse, and yet even with such sacrifice, on average only two out of several thousand young survive to adulthood.

Octopuses are an ancient lineage, and quite unique amongst invertebrates for their high level of intelligence. They are deeply emotional animals and loving parents, but climate change could spell an end for them. Whilst deep-sea dwellers like the Dumbo octopus are safe for now, most other species live in shallow surface waters that place them at risk. Climate change results in ocean acidification, which impairs the ability of octopuses to draw oxygen from the water. That's a major issue, with vision impairment and eventual blindness two of the most immediate consequences. Cuisine sees octopuses served up as a delicacy in the Mediterranean and some Asian regions. This decimates local populations and abuses captive animals in octopus farms. The celebrity-physicist Brian Cox amongst others has refused to eat the creature, having witnessed first-hand the remarkable mental capacities they show.

Joining my ark, the dumbo octopus serves as a case in point that this Earth is richer than we know- countless species are waiting to be discovered boasting all manner of superpowers. However, it also shows that the harm we humans have caused is indiscriminate in the animals it affects, bearing down on even the hardiest survivors and Einsteins of the natural world.

5.) Tapanuli Orangutan (Critically Endangered)

I've always wanted to live in a treehouse. The ability to look down at nature with a bird's eye view seems so magical to me that it has never lost its appeal. Orangutans not only fulfill this fantasy of mine, but they do so in spectacular fashion. Say I was to drop you in the rainforest canopy of Indonesia to spend the night. My guess is, you'd get precious little sleep and possibly even break your neck with a 30m fall to the forest floor. Orangutans, on the other hand, have evolved to spend most of their life in this complex 3D landscape, building beds for themselves in the trees every night as they settle down to rest. In a little over the time it takes you to brush your teeth, they construct their nightly nest, often replete with pillows, blankets, a roof and/ bunk-bed.

Whilst chimpanzees and gorillas are easy to study, orangutans remain just slightly out of reach, high up in the treetops on remote leech-infested islands. Biruté Galdikas was the one who sought to change that. Alongside Jane Goodall studying chimps and Dian Fossey with her gorillas, Biruté was the third of the 'trimates'- female researchers who revolutionized our understanding of our closest relatives. She first entered the Bornean rainforest in 1971 and has since continued to work with the great red apes for a full fifty years. Through rehabilitation and wild observation of orangutans, Biruté has made the single largest contribution to our knowledge of the species, leading one of the longest continuous studies of any species.

In late 2013, an injured male orangutan named Raya was brought to the researchers. Bearing cut wounds and marks from air rifle pellets, Raya was in a very bad state, having clearly been harassed by local villagers. Despite intensive veterinary care, Raya died just eight days later. Several years passed and new evidence came to light in the form of genetic studies. What was previously thought to be a single orangutan species isolated on the island of Sumatra was in fact two separate species inhabiting the North and South parts of the island respectively. Raya's skeleton was re-examined and became the first-ever Tapanuli orangutan described.

With the addition of this third orangutan species, the rarest great ape on the planet was born. The Tapanuli orangutan may number as few as 800 individuals, though counting them in the dense jungle is a challenge. An unusual partnership has seen astrophysicists and ecologists unite, with thermal imaging drones running technology intended for recording stars now used to map the heat signatures of orangutans in the treetops. The habitat available to all orangutans is severely fragmented from palm oil plantations and community settlements, so knowing whereabouts they are is more critical than ever. These are the smartest non-human apes, the loss of which would be sorely felt.

That's where zoos come in. Orangutans aren't especially common in zoos across the globe, but those few in captivity are important as safeguards should their wild cousins disappear. There are always questions raised about keeping highly intelligent animals in zoos, but I personally come down strongly in favour of the practice. It only takes a quick look at the zoo facilities in question to see why. In Milwaukee Zoo, the orangutans have each been given an iPad loaded with games, David Attenborough documentaries, and Skye so they can call orangutans in other zoos. In the Netherlands, one zoo found their orangutans had lost some agility at swinging through trees, and so brought in an Olympic gymnast to teach them the ropes. Though the apes weren't all that impressed by his antics on their newly renovated climbing frame, it gave the public quite a spectacle to behold. If I were an orangutan, I'd much rather be in a nicely kept zoo exhibit than in the wild.

Orangutan is a word meaning "man of the forest," and their behaviour leaves me with no doubt as to their strong human resemblance. Researchers working with them in the wild have testified to seeing them mimicking all manner of actions. Sawing wood, applying insect repellent, using soap and washing clothes, using hammers and stealing boats all count on a long list of orangutan escapades. They use a paste made from pulping up a particular plant species to soothe muscle aches, and it is thought that local tribes learnt from watching orangutans how to make use of this natural remedy.

The Tapanuli orangutan joins my ark in recognition of the fact that we're far more alike to most animals than we tend to give them credit for. And that there are some skills we cannot hope to match other animals at- like building treehouses.

6.) Komodo Dragon (Vulnerable)

Indonesia is one of the most biodiverse countries on Earth, so it's hardly surprising we should find Earth's most impressive reptile amidst its species-rich islands. Reaching 3 metres in length and weighing as much as you or I, the Komodo dragon is as close as we've come to finding a living dinosaur. They can smell blood up to 4km away, and inject 5 separate toxins in their venom when they bite. Victims experience painful cramping, loss of body heat, blood thinning, heightened pain sensitivity, excessive bleeding, shock and unconsciousness in no particular order. So it's no great surprise that we're wary of the creatures.

In 2013, a Komodo dragon walked into a ranger's hut. What ensued was not a joke. The people inside were caught completely off guard, and the dragon promptly bit the ranger and another member of staff before the situation could be brought under control. Both were evacuated to hospital for treatment, though fatalities are rare. Since records started in 1974, over thirty attacks on people by the dragons have been recorded, five of them proving fatal. And with the tourism industry booming off the back of our morbid attraction to the creatures, habituated Komodo dragons are becoming an ever-greater threat.

They are a perfect example of why conservation is so hard.

Local communities have lived with the dragons for centuries, learning to coexist and fending away the reptiles with nothing more than sturdy forked sticks. Dragons naturally hunt deer and the occasional water buffalo, so humans were a fleetingly rare feature on the menu. Komodo dragons were everywhere but people were few and far between, so conflict could be avoided. In the early 20th century, Europeans discovered the dragons for the first time, and it didn't take long for the trickle of visiting naturalists to turn into a flood of excited foreigners vying for a glimpse of the beasts. Recently the local airport was upgraded, with its capacity increasing tenfold to allow 1.5 million visitors annually to the region.

On the face of it, this shouldn't have been too difficult to manage. Rare cases of dangerous dragons known for attacking humans were countered by relocating the problem lizards to a remote, smaller island away from the main island of Komodo. Except that tourists have started to go further afield, and are now using this island too. It's a bit like Jurassic Park being rediscovered by a group on a school trip- something's bound to go wrong.

The area is a World Heritage Site, placing the local government under extra pressure to ensure it's well conserved. And long before COVID came along, plans were formed to close the main island of Komodo to tourists for 2020. After much to and fro, the plans were dropped. Local communities were desperate for the revenue tourism brought; dragon populations overall seemed relatively stable; and, though prey items such as deer were in decline, more concerted efforts to monitor and manage their population could easily remedy the shortage.

What the right answer might be, nobody can say. But the complexity of the situation, balancing human and wildlife needs, is common across modern environmental management practice. And there's never just one problem to look at. A growing threat from the illegal wildlife trade has been noted, with as many as 41 Komodo dragons being smuggled through transactions on Facebook. How they were caught and shipped off is anyone's guess, but each of the dragons sold for over $30,000. Authorities must adapt constantly to meet these changing threats, and the current dragon population of several thousand is a testament to their success.

The Komodo dragon aboard my ark attests to the trying complexity of modern conservation, and to the awe we feel towards other living creatures. Protecting such animals results in far-reaching benefits for all involved, provided they can juggle both respect for nature's power and tenderness in accomodating it.

7.) Magnificent Sundew (Critically Endagered)

I'm going to say right away that I am not a plant person. Tree-hugger perhaps, but if it doesn't have eyes or a heart or a nose, then it's off the books for me. I'm not proud of my anti-plant bias, but the reason for it is simple to devine. Plants don't move or do anything much interesting on time scales we can appreciate. The exception being carnivorous plants, some of nature's deadliest hunters, amongst which the sundew counts itself.

To picture a sundew, imagine a cartoon octopus with flailing legs. Turn those legs into bright red stems, and add thousands of tiny droplets of sticky fluid to the broad leaves. It's a start. Basically, unwary insects fly into the sundew's coated stems, and the plant grabs hold of them tightly whilst digesting them alive. That's not a nice way to die, even if you get to be eaten by a beautiful plant.

It was a picture of the aptly-named magnificent sundew that caused it to be found; a picture on Facebook of all places. It started when a plant enthusiast living in a remote mountainous region of Brazil went for a walk. On this particular walk, he came across a very attractive plant that he took some pictures of and posted onto social media. One year later, and through a long bread-crumb trail of sharing on Facebook and via friends of friends, these images happened to arrive in front of a young researcher in the town of São Paulo nearby. They appeared different from any of the sundews he had previously seen, so he set out on an expedition to meet with the enthusiast who had found it and scientifically describe the finding. It was a new species, completely unknown to science, found thanks to the worldwide web.

That very same year, another new species- a type of flying green insect called a lacewing- was found by a scientist trawling through photos on Flickr. This is so far removed from our sense of high-brow, rigorous academia that it's almost comical. But the fact remains that scientists are increasingly reliant on everyday people exploring and posting their findings to improve our knowledge. The magnificent sundew was found on a single mountain peak at 1500m elevation. Its habitat is completely surrounded by cattle ranches and plantations. The odds were very much against its discovery, and still remain stacked against its survival. Yet with the knowledge that it exists, there's a slim chance that we're able to conserve it one way or another.

Sundews famously live in some of the least plant-friendly soils on Earth, gaining the essential nutrients they require to survive by eating small insects. They're successful simply because they have little or no competition, similar to how ravens are able to lord it over tips and rubbish heaps. The remarkable thing about sundews, however, is this ability of theirs to react in human time to their surroundings. These are predatory plants, and just as cool as that title makes them sound. We've even discovered one sundew species that reverse-engineers its insect-snatching mechanisms to alarm its flowers against hungry moths that could damage them. The slightest touch and the flower closes, leaving it wrapped up in its petals, safe from harm.

Flies live life in the fast lane with a higher metabolism that means they perceive time passing in a sort of slow-motion. That's why the inane buzzing critters are so hard to swat. Plants have low metabolisms, live longer, but also perceive time as fast-forwarded as a result. So whilst you need bucketloads of patience to watch a tree growing, the tree sees itself shooting out of the ground much faster. That fact alone is enough to win a plant a spot on my list.

In Melbourne, Australia, 70,000 trees were each given separate email addresses to inspire the public to connect with the greenery. The results were touching, and popular trees received love letters from admirers across the globe. "You are the gift that keeps on giving," said one. Another continued: "I was struck, not by a branch, but by your radiant beauty." From another: "You inspire me to live life to the fullest, and pursue my dreams." Plants, on the whole, are hard to fully appreciate, but plants that respond to your emails... well, that's a little easier.

The magnificent sundew joins my ark as a statement that the plant kingdom ought never be forgotten: likewise for the fungi, bacteria and other microscopic life that struggles on all around us. Taking the time to look a little closer can lead to unexpected breakthroughs, and technology can be a potent force for good.

8.) Española Giant Tortoise (Vulnerable)

We've brushed shells with one of the Galapagos' giant tortoise species- the Fernandina tortoise- already. But aside from that passing mention in chapter 3, these incredible animals have thus far gone unmentioned. No longer. Giant tortoises, of course, helped Darwin to arrive at his grand theory of evolution, though they get precious little in the way of thanks. None of the dozen varieties found on the archipelago were described at all scientifically for 300 years after they were found. The reason comes down to the remarkably good eating they made, praised by sailors for their delectable taste such that none of the creatures loaded onto ships ever reached port in London.

Since marine voyagers first chanced upon the island in 1535, it became a regular stopover for vessels looking to replenish their stores. Tortoises were loaded onboard and stored alive below deck, stacked together and immobile. Kept as such, the meat would last for months, and came conveniently with a large store of fresh water from the tortoises' bladders. Iron-ically, the hardiness that had helped them to survive, going for six months without food or water, and floating between islands, made them ideal for larders as well. Darwin himself is due his fair share of the blame. A naturalist with a taste for the exotic, his ship carried off a whopping 700 tortoises when they passed through the Galapagos in the 1800s. The species they abducted, the Floreana tortoise, went extinct only a decade later.

So afflicted with extinctions taking place left, right and center, giant tortoise species have become somewhat of a symbol for conservation globally. When Lonesome George, the last Pinta Island Tortoise, died in captivity in 2012, the whole world mourned his loss. And it is in the face of such adverse circumstance that the Española tortoise shines through as a ray of hope.

By the 1960s, the population of tortoises on Española was at an all-time low: 12 females and 2 males rounding out the entirety of island's giant reptiles. In a brave move, authorities decided to bring in every one of the tortoises and relocate them to a captive breeding facility. Should this have been unsuccessful, there would have been no end of blame levelled against them. Getting animals to reproduce isn't as easy as it sounds. But thanks to the iron temperament of a male called Diego, as many as 800 tortoises have now been sired in his line. That's a feat which the sexually prodigious Genghis Khan himself would be jealous of! Starting in 1975, tortoise reintroductions to the island have added some 2000 animals. The permanent population on Española is stable at around one thousand giant tortoises, and threats from invasive rats are being brought under control.

"Phew- that was close" is a phrase conservation essentially invented. Countless species have been brought back from the brink, and the larger the animal, the more attention it attracts- for better or for worse. The Galapagos islands are world-famous as an ecotourism destination, with waves of tourists washing up on their shores. In the past decade, two hundred hotels have sprung up on the islands, catering for the hundreds of thousands of visitors who pass by each year. Land-based tourism has doubled, and whilst marine traffic is strictly regulated, few rules apply on terra firma. Tourist activities bring in revenue which funds the world-class environmental protection the archipelago sports. However, if unchecked, this lifeline could also ruin the wildlife that creates the attraction in the first place.

The Española giant tortoise joins me on the ark to bear witness to the fact that courageous acts are the lifeblood of conservation. We as humans have the potential not only to radically harm ecosystems, but to radically improve them. I know which option I'd prefer.

9.) Wallace's Giant Bee (Vulnerable)

Alfred Russel Wallace was a mirror image of Charles Darwin: 15 years his junior, but still white-haired, long-bearded and obsessed with notions of evolution. The differences between the two were small: Wallace wasn't bald on top and was obsessed with insects, whereas Darwin was a barnacle fanatic. The two of them came together in 1858 in front of a small group of white-haired, bearded, balding British scientists who had gathered to hear them speak.

Wallace had spent many years travelling around the tropics, collecting and observing the creatures that inhabited the region's steamy rainforests. In doing so, he came to a remarkable realization. A realization that, by chance, had occurred to Darwin himself in his travels, though which he had never published writings on. The notion that had occurred to them both was that of natural selection, the driving force behind evolution. Wallace was a great admirer of Darwin, and wrote to him with these thoughts. They co-presented their research, but Darwin's name lived on in the history books as the great mind who discovered evolution, whilst Wallace was largely forgotten.

Except by scientists, it seems. For despite wearing opaque safety glasses and ridiculous face-shields, scientists pride themselves on attention to detail. And seeing how unfairly Wallace's name was dropped from public knowledge, they've taken it upon themselves to settle the score. Alfred Wallace has a grand total of 390 species bearing his name. Darwin himself has under 300. Whilst Wallace's passion for insects has caused scores of them to be named after him, Darwin has his name on only one species of barnacle. To mark the latter's 200th anniversary, a parasite found in the intestines of the short-nosed bandicoot was named in his honour. Revenge is sweet.

As for Wallace's giant bee, the name says it all. This mega buzzing bulldozer of an animal is five times the size of your common-or-garden honeybee, as large as your thumb. Also unsurprisingly, it was first discovered by Alfred Wallace in his jaunt around the Malay Archipelago. Then it went missing for a century. Then a researcher sighted it briefly, and it blinked out again. Size is everything, and as we learnt in chaper 3, there have been far larger creatures that managed to slip under our radars. A few years back, when the Global Wildlife Conservation charity set up the 'Search for Lost Species' campaign, Wallace's giant bee featured on their hitlist. The hunt was on.

In the dawning new year of 2019, a team set out to find the missing mega-bee. The location and timing of their search were carefully chosen to coincide with previous sightings, but despite their preparation and expert local guidance, the species was a no-show. One member succumbed to illness and had to be evacuated. Those that stayed behind scoured two neighbouring islands, marking and guarding the tree-dwelling termite mounds in which Wallace's giant bee was known to burrow. Still no results. Finally, drained and defeated, the diminished team set off on the return trip. Walking with the forest on one side and an orchard on the other, they stumbled upon a termite mound with a bee-shaped hole in it. Wallace's giant bee was not lost after all.

There's somewhat of a love-hate relationship between people and bees. Some people fear them, being allergic to or just plain terrified of their sting. Others think they're worth their weight in gold. They're immensely valuable to farmers in pollinating crops, aside from producing delicious honey. Historically across Europe and the US, households that kept bees would inform them of any major news, and some would be invited to funerals. They've been taken to space (flying well in zero gravity), trained to sniff out drugs and explosives, even set aside designated 'highways' of flowers to travel across cities. McDonalds made a miniaturized restaurant that functions as a bee-hive and a suite of their stores have continued the tradition with rooftop instalments.

It's not all good news. Colony Collapse Disorder became a global concern around the turn of the century with bee swarms increasingly and inexplicably abandoning their hives. Speculation suggests anything from pesticides, mites and fungi to malnutrition and pathogens to beekeeping practices like antibiotic use and long-distance transportation may be responsible. Nobody knows, and the chances are that many or all of these factors play a role. A class of pesticides called Neonicotinoids have attracted widespread controversy when it was discovered they were toxic to pollinating bees. The chemicals have been banned across many countries, but mass insecticide usage still takes a large toll on wild populations.

In honour of the mixed relations humans share with all sorts of creatures, Wallace's giant bee joins me on my ark. Often discoveries pop up in the most unexpected places, as nature's way of keeping us all on our toes. When a number of beehives in France started producing rainbow honey, few would have suspected a nearby M&M plant was to blame. And when we've given up hope, nature can still surprise us as it did with the reappearance of Wallace's giant bee.

10.) European Turtle-Dove (Vulnerable)

Titan is an unusually masculine name for a dove, but Titan he was called. Carrying a satellite tag with him on his month-long marathon from Britain to North Africa, this athlete of a bird was recorded flying over 500km each night at a speed of 60 kilometres per hour. Past the Atlas Mountains and the Mediterranean Sea, across the Sahara Desert he flew- only to reverse the journey 6 months later, returning to the UK to breed.

Migrating birds are one of those things that scientists just couldn't work out for centuries. You might remember Aristotle in chapter 1 saying that some birds hibernate, and others transmogrify into different species with the changing seasons. Those ideas seemed sensible enough at the time, so continued to be believed. Come the 1600s, and another great thinker- a certain Charles Morton- came up with his own radical explanation. On scant evidence, he claimed that contrary to Aristotle's musings, what birds really do is fly to the moon. This wasn't just a wild claim: he set out the trip in great detail, taking 60 days on the wing at the alarming speed of 200km/h. He knew that at higher altitudes there would be less air to provide friction and less gravity pulling the birds to Earth, which seemed to support his idea. He reasoned that they could burn through excess fat reserves so as not to require food on the journey, and that they could sleep on the wing.

A nice thought, certainly, but sadly not the case. Birds do burn through fat reserves and some do sleep on the wing. However, few of them can fly at 200km/h for any length of time, and that's 1/100th the speed required for them to escape from Earth's orbit. No bird could survive in the vacuum of space, and honestly I don't think any would be mad enough to try.

In 1822, a strange bird arrived in the town of Mecklenburg in Germany. This was a white stork like hundreds of others that came to the area at that time each year. What made it strange was the addition of an arrow sticking out of its neck. It was an age of guns and firepower, so the foreign arrow really was most peculiar. The wood it was made from wasn't found in Germany, and the wound it left looked weeks or months old. The only realistic explanation was that the bird had been shot whilst travelling through Africa. Evidence was mounting.

In 1899, a Danish school teacher came up with a clever idea to test where birds went when they disappeared. Hans Mortensen caught all sorts of birds from starlings to storks, herons to gulls to ducks. He would attach a small metal ring around one of their legs inscribed with an address and an identifying number. Anyone that found a bird with one of these rings was instructed to send him a letter containing the band's number, along with when and where it was found. All of these metal bands he made himself from an aluminium sheet, and he personally banded

6000 birds in his lifetime. The outcome was not only proof of long-distance bird migration routes, but the birth of a new scientific discipline in bird-banding.

Nowadays, this practice continues very little changed from Mortensen's original methods. The data is collected centrally and used to find out all sorts of useful information about birds, from behaviour to breeding cycles, movements to moulting times. The satellite tag fixed to Titan was a high-tech alternative to meticulous handwritten records, precisely mapping his location at regular intervals throughout the day and night.

The trouble with birds making such long-distance hauls is that they are subject to the whims of every country they pass through. It's a long chain, and a single weak link can collapse entire populations. European turtle doves like Titan are in particular trouble. Their numbers have declined by 96% since 1970, and a large part of the reason lies in the mid-way point of the Mediterranean that they cross on their migration. Countries like Egypt, Italy and Lebanon all take their toll, though thankfully legislation is being tightened to impose harsher penalties on anyone caught hunting the birds.

The black sheep in this instance is the tiny island of Malta: the only country in the EU to allow spring hunting for recreation. 10,000 hunters engage in the practice, with a quota of 11,000 doves which can be legally shot. The true death count is hard to determine, but the impact is disastrous not only for doves but for countless other threatened species illegally taken out by trigger-happy hunters. A 2015 national referendum on the matter voted narrowly to continue the practice, a mere 0.4% majority seeing the slaughter continue.

The European turtle dove rounds out my ark's cargo, as a reminder that even today there are huge discrepancies between cultures and each countries' commitment to conservation. Protecting migrating species requires absolute trust and collaboration between national authorities, which is sometimes too much to ask. If you're after a happier note, then we'll pick up chapter 7 there, looking at the most inspiring success stories and innovative solutions being put to use in saving our species.

7

Small Packages

My entire life fits in the space of five minutes and twenty seconds. That's not the result of my time-travelling antics, but rather a worrying note on the mounting threats we as humans face today. In 1947, amidst the dying strains of World War Two, a group from the Bulletin of Atomic Scientists convened and came up with an idea. As a symbol of how close we were to disaster, they would show a clock face depicting a certain number of minutes to midnight. If it ever reached midnight, then we'd be in the middle of global catastrophe. Happy thought, isn't it? At the time, the hands were placed at 7 minutes from the hour, and in the years that followed its minute hand swung back and forth from 2 minutes in 1953 to a full 17 minutes in 1991 and so on. Right now we're at the latest hour we have ever been- a mere 100 seconds from the end of humanity.

Should we be worried? Probably. Does that mean we shouldn't act? Of course not. And herein the age-old paradox of environmentalism lies. Say I told you that tiger numbers were on the rise for the first time in a century- that's great news, right? Does that mean we should focus our attention now on saving other species? Certainly not. Just because there are more tigers now than five years ago doesn't mean they have recovered. Their numbers can be bad but getting better, can they not?

There are the most amazing, dedicated, inspiring conservationists who've orchestrated this turnaround. Looking at the absolutes of 'good' and 'bad' means that relative terms defining their success- 'better' not 'worse'- get swept underfoot. I've already shown you that things are bad; this chapter is about how they're also getting better. We'll take a possibilistic stance, seeing what could be, as opposed to what is or has been. So with that said, let's hit the road. First stop Mauritius.

In 1974, the Mauritius kestrel was the rarest bird in the world. Four of them remained on the island, the rest having fallen prey to a smorgasbord of threats. Over three centuries, the infamous Dutch sailors that eradicated the Dodo made the island hell for the rest of its inhabitants. All but 2% of the native forest was cleared; cats, mongooses and crab-eating macaques were introduced; and a malarial outbreak in the 1950s and -60s caused the toxic insecticide DDT to be sprayed across all surfaces, killing off many smaller lifeforms. The Réunion kestrel on a neighbouring island was extinct by 1700. The Mauritius kestrel wasn't far behind.

In short, most scientists thought of any attempt to save this species as a joke. And rightfully so: the odds of success were a thousand to one. But a group led by our old friend Gerald Durrell set about the mammoth task. They created a sanctuary on a tiny coral island offshore. They also attempted captive breeding the kestrels, but the incubator broke down and killed the chick inside. With only a single breeding female, the species' existence rested on a knife-edge. Many of the eggs laid were infertile. Yet one by one the numbers climbed. Young kestrels formed their own breeding pairs, and by 1984, the population totalled 50 of the birds. The team could now take and rear eggs from wild kestrels' nests, allowing the female to lay again and rear her second egg unimpeded. At the latest count, 400 Mauritius kestrels were recorded, making this one of the greatest success stories of all time.

The unlikely comeback of the Mauritius kestrel is heart-warming to read about in hindsight, but such a precarious situation must have been terrifying for those working to solve it. Mauritius remains the deadliest place on Earth for animal species, given a survivability index of 0.39 by the IUCN. Sri Lanka, the country with the second-worst score, sits at 0.56 out of 1. A measly 0.001% of Mauritius' lands and oceans are protected, representing an almost insignificant contribution towards conservation. We can celebrate successes like that of the Mauritius kestrel whilst still recognizing that underlying threats remain. And nowhere is this philosophy more critical than in animal rescuing.

Nuisance. That's his name- Nuisance. Outside my room, I can just about see through my window the lorikeet called Nuisance. I rescue injured wildlife here in Sydney, and lorikeets come up more often than most. After a while, you run out of half-decent names for them all. And besides, I think he quite likes it. Nuisance is kept company by a cockatoo thrice his size, both of them part of a small, ever-changing menagerie of animals I care for. I've given flying lessons to owls, made bandicoots porridge, been pissed on by possums, and spent many agonizing minutes holding lorikeets whilst they swivel their heads 180 degrees and gauge large chunks

from my fingers.   Herbivore my foot!   At least when I go
prematurely deaf, I'll have a small list of birds to blame. Oh...
and incidentally, if you're ever looking for a replacement
fire alarm, a lorikeet strapped to the ceiling would outshout
anything on the market.

Jokes aside, I love these animals. They are resilient and
charismatic and have enriched my life no end. And their stories
are made all the sweeter by the public goodwill on which the
rescue system depends. People from every walk of life seek
help for wild animals in distress, allowing trained volunteers
like myself to reach the scene in good time. I've met those
who've postponed travel plans, taken work leave or stayed up
late into the night to see that the animal they've found is safe.

Now, I can complain about Australia's conservation negligence until the sun goes down, but to have a system like this up
and running is superb. It means there's a tremendous portion
of the public who really value their native wildlife, and will go
to great lengths to see that it survives. That gives me hope.

I had a duckling in care recently for six weeks, and the sheer
joy she displayed in her first swimming lessons had to be seen
to be believed. I would get into the pool and swim ahead of her
with her cuddly toy duck friend. The duckling, whose name
was Puddles, followed obligingly behind. Within days, Puddles
was faster than me and would add dives and wing flaps to the
routine. When she was returned to her grassy pen, she'd wait
until I was out of sight then break out and return to the pool for
some extra swimming time. I'm not a sentimental person, but
Puddles knew how to pluck at the heartstrings. She earned the
nickname 'Princess' for being spoilt, chaperoned and doted on
endlessly. I don't think she was looking forward to her release.

Another recent rescue case was an echidna (named Edna)

under canine attack in one of the neighbourhood's backyards. The dog safely restrained, it remained only to relocate the spiny monotreme to a nearby national park. For the clueless readers, an echidna is like a giant hedgehog the size and shape of a large bowling ball that's been partially flattened by a road train. And boy are they strong! With it firmly entrenched in a small hole it had dug, I was forced to move from digging with bare hands to a trowel and eventually a full-blown spade to get it out. The one thing you must never doubt in nature is its stubbornness to survive.

Zoos are sometimes seen as a controversial point, and certainly there are some which deserve to be shut down. But for the most part, the services they provide are invaluable for education, science, conservation and animal welfare. I've mentioned Gerald Durrell a few times before, and I might as well admit he's an all-time hero of mine. Part of an older generation of collector-naturalists, Durrell wrote some wonderful books about his adventures and shaped the role of the modern zoo in conserving threatened species. A good criterion for judging a person is the number of pictures of them with animals you can find, and on that count, Durrell is head and shoulders above the pack. Anyway, part of Durrell's legacy lies in his pioneering use of zoos as breeding centres for endangered animals to be reintroduced into the wild. The Tasmanian devil shows precisely why that is necessary.

Famously caricaturised in Taz from the Looney Tunes, the Tasmanian devil looks rather like a cross between a pit bull and a sun bear. They're the primary scavenger on this Australian island state, disposing of carcasses to stop the spread of disease through the ecosystem. In the mid-1990s, reports emerged of a facial tumour disease amongst the devils: a sort of infectious cancer passed between animals during their aggressive mating encounters. What was once a population of 140,000 dropped sevenfold in the space of two decades, disrupting the food chains they were a part of. Authorities began to get desperate, and desperate times call for desperate measures.

Starting in 2005, approximately 130 Tasmanian devils were taken from the wild and placed in zoos across Australia. Some made it as far as New Zealand and the US. These disease-free animals formed the base of a new stable colony now numbering over 500 animals in 26 institutions. Meanwhile, researchers set to work developing a vaccine for the facial tumour disease, and a few years back they succeeded. Now it's a matter of returning captive-bred animals to Tasmania, fully-vaccinated. They'll be able to field-test the cure and carefully monitor these animals to ensure the devils' resistance to the cancer continues. Thanks to the network of devils in zoos, a large amount of the species' genetic diversity has been kept in spite of wild populations continuing to fall. So zoos saved the day, as they so often do.

Which brings me in a roundabout manner to an all-time favourite story of mine. It's about a very special echidna named Matilda and her struggles. Matilda was a tough critter: orphaned at a young age, she was brought to a wildlife sanctuary where she spent the first few years of her life plagued by chronic medical conditions. She had constant discharge from her eyes, a red-raw underbelly and severe hair loss. The flummoxed vets tried every known treatment to cure her. She was given regular rounds of antibiotics and routinely bathed, provided with ointments, salves and topical cures of all descriptions. But none of it had the slightest effect on poor Matilda. Finally, they opted for an allergen test, using a panel of all the things she might encounter in her daily routine.

It turns out that Matilda was a sensitive creature. She was allergic to many of the common native plants but most surprising of all, she was also allergic to ants! Now, ants form pretty much the entirety of an echidna's natural diet, so this was quite a major issue. Imagine being lactose intolerant, suffering from celiac disease, bearing allergies to soy, shellfish, nuts, fruit, fish and veggies and you still don't come close to the equivalent. 'Incredible' springs to mind as a word for Matilda's perseverance, and with the correct diagnosis made she was rapidly treated with a special vaccine designed to build up her immunity.

For many of us, stories like this are far removed from the daily routine. Zookeepers are a small elite, and wildlife carers are mostly grey-haired retirees. What does conservation mean to your average Joe or Joeline? Perhaps not much. But now more than ever, there is opportunity for more people to get in on the action. That seems contradictory because we're moving into big shiny cities and away from big dirty forests, distancing ourselves from the natural world. Sure, we're continuing to create environmental problems, but you can't save lions from a London flat. Correction: you can. Thanks to citizen science.

When I tell you the term 'citizen science' entered the Oxford English Dictionary in 2014, you should be impressed. That means this idea is cutting-edge, but actually it's also pretty darn intuitive. People are interested in the world around them- that's why you're reading this book. But people like doing things, not just reading about them. So if you give people interesting things to do, then you can get a lot done very quickly. Many hands make light work. I told you it wasn't hard to grasp.

Simple idea it may be, but citizen science is powerful. Right now, iNaturalist is leading the charge, and the platform's figures are there to prove it: 900,000 observers, 100,000 identifiers, and 33.5 million records made of animals across the globe. People all around the world are taking photos of creatures, uploading them to the database, and (here's the exciting part) having their findings confirmed and identified by thousands upon thousands of experts and enthusiasts. Surveying for biodiversity isn't a new concept, but collecting data on this scale is unprecedented. We live in a world that runs off big data, so it makes sense to gather big data about the world we live in. Every photo observation is verified multiple times and graded for research purposes via the consensus that identifiers reach. That means serious scientists can work with it, policies can be written with this knowledge and goodness knows what else.

The less techy version of this is found in bioblitzes. Bioblitz is a cool word, but its meaning is even better- an event where ordinary people get together and work over a day or a weekend to find as many species as possible in a given area. It's competitive science that anyone can be a part of. For the armchair traveller, there are projects like Snapshot Safari where you can surf through images uploaded from hundreds of remote cameras across Africa. Participants are invited to tag the images with the animal species in the frame. Again, no skill whatsoever is required- if you can tell the difference between a rhino and a rock, then you're qualified. Although that can be quite a challenge, from experience!

Citizen science is valuable, worth several billion dollars annually, and certainly not exclusive to environmental pursuits. Platforms like Zooniverse compile these projects to let users explore and find something that tickles their fancy.

Astronomers and historians are just as keen for public input if those fields are more your cup of tea.

Yet the value citizen science provides is more than the sum of these helping hands. It's connecting people in an unprecedented way to the world around them, democratizing nature appreciation. Reality TV appeals to us because it's 'genuine', and whilst I can't say I've watched an episode of the Kardashians, I'm pretty confident in saying that life itself is the ultimate reality show. With just a few clicks I can livestream the action from a busy African waterhole, or peek into a condor nest, or a panda enclosure, an aquarium or an elephant sanctuary. Nature has never been closer.

So, we've looked at people-powered solutions, and at the people-technology intersection. What about technology itself? Remember the Interspecies Internet idea from chapter 5? It's rapidly becoming a reality. In the past decade, we've developed mind-blowing technological systems to decode ancient human languages. By analysing inputs from texts on tablets and other artifacts, beautiful cloud diagrams are made showing the linkages between all the language's words in hundreds of dimensions. If you can't imagine what hundreds of dimensions look like, neither can I! However, I feel I'm safe in saying it is complicated.

When you use a dictionary, say between English and French, you're looking up a word that matches your intended usage. 'Potato' and 'Pomme de terre' are equivalent words in that they share the same linkages in sentence construction. Because finding the right set of linkages equate to translation, these computer-generated diagrams can be overlaid, with matching intersections providing translations for thousands of words. Voila! It's the technological equivalent of moving from the

Rosetta Stone to Google Translate. And there's a very real possibility of decoding animal languages with it.

Talking with animals aside, there are some easy steps that can be made towards communicating and collaborating with other creatures. The golden rule is that you have to give the animal(s) in question agency. Agency to alter their interactions with humans and agency to change their environment.

Sampson is a Macaw at San Diego Zoo. Sampson loves music. So, working with MIT, the zoo have made Sampson a very special toy: a music playlist controlled by one of his perches. A jutting branch has been rigged up to act as a switch, such that by holding it down with his foot, he can listen to all the latest hits. Often he'll dance along to the beat by bobbing up and down on the perch. Hence the second iteration of the toy added a motion detector that keeps the music playing for as long as he keeps dancing. The technology isn't particularly incredible, but the thinking behind it is brilliant in keeping this intelligent bird happy and occupied. Plus the zoo visitors love it!

Then there's the Elvis Project. This involves a screen set up in a dolphin enclosure with 16 hydrophones in a grid pattern behind it. They sense the intensity of the dolphin's echolocation and light up a back panel to show the hot spots. The keeper can use this visual cue to know where the dolphin's echolocation beam is pointing. And the upside for the dolphins is they can order room service, or tank service rather. The dolphins are provided with their own underwater menus- a choice of fish- and know to point their echolocation at their item of choice. They then swim over to their keeper who is more than happy to oblige. The next step for the technology is the D-pad, this one an infrared system that detects the position of a dolphin in front of it. So a few lucky dolphins get to play on special dolphin apps, with games like "Whack-a-fish" to amuse themselves with.

Whimsical invention or world-changing idea? At this point, it's rather hard to say. There are benefits right now provided by these systems, notably as enrichment for animals in captivity. But beyond that, any further impact will depend on how far the idea can be stretched. If we could hold full conversations with dolphins, that would be a major breakthrough. With learnings for both sides, no doubt.

Ten years ago, drones were just arriving on the scene. The technology was still in its early days but had advanced sufficiently that the price-tag, size and performance were practical for more widespread applications. So emerged a new chapter in the anti-poaching scene- national parks across Africa leapt on this silver bullet solution to tracking poachers across vast terrain. The unmanned aircraft were equipped with thermal imaging and night vision technology to better spot their targets at altitude. Any sign of poachers and an alert would be triggered to send out a poaching patrol on the ground. There was great excitement amongst authorities that the war on poaching might finally be won. It wasn't.

Drones at the time had insufficient battery life, poor handling in adverse weather and a suite of other flaws that saw their usage in anti-poaching crash and burn. National parks were faced with ongoing repair costs to use the valuable hardwire, and in the end, it was dropped from the register.

Nowadays, drones are being used to great effect for this exact same purpose. That short space of time has made all the difference, and slowly the threat is being brought under control. Rhino poaching in South Africa has less than halved in five years, as you know from the Preface quiz. These new drones are being trained on computer-generated models of the landscapes they'll be operating in, including plants, people, and animals. The AI detection software they use to determine which far-off shapes are people and which are prides of lions stands up to the test in all conditions. Thus rangers can be deployed with confidence and precision where threats arise.

There are other tools at the disposal of protected areas to improve the work they do. Take "Earth Ranger," which uses machine learning to predict animal movements and forewarn of possible human-wildlife conflicts. Or the SMART system, that computes the best anti-poaching patrol routes to take, based on animal tracking and previous poaching threats. WildMe has developed programmes to identify animal species from footage or photos, and then taken this one step further in identifying individual whales, whale sharks and manta rays. Being able to ID individual animals is a huge benefit for those working to conserve these ocean giants, because their numbers are relatively small and each individual's moves important.

A different tack is taken by the Satellite Stories project created around Ol Pejeta Conservancy in Kenya. Mixing animals' tracking data with computer-simulated movements, they created a hybrid of Google Maps and Attenborough-esque storytelling in website form. The central page shows the movements of hundreds of animals across the reserve over the course of a week. A choose-your-own adventure function lets you follow narrated story paths illustrated by the moving animal dots, or you can browse the unfolding scene at your leisure. This is a sort of nature documentary on a much more personal level, and with the scope to place the viewer in the director's seat.

At the most basic level, even being able to show more animals to more people is important. Joel Sartore is a National Geographic photographer with a mission. He has set himself 25 years to photograph all of the species in human care around the world. The Photo Ark, as the project is known, contains over 30,000 pictures of 10,000+ species to date. These portraits are a unique portal for people all across the world to witness life's diversity, and to act as a time capsule should any of these species disappear. It might not be what we commonly consider as high-tech, but with his camera and a stack of memory cards, Joel Sartore has created a unique archive of Earth's biodiversity.

Nowadays, we're in a Cambrian age for technology, just like the explosion of life forms that occurred at this time 540 million years ago. As whole new fields of pursuit emerge, each is finding an application in solving today's environmental issues. The opportunities are truly limitless. Yet whilst the Cambrian spawned sea monsters, our technological equivalent isn't quite as sexy. So I'll confine myself to just one further example that stuffs in as much technology as could be imagined.

With a face like a mildly confounded angel and an utter fearlessness amongst humans, the New Zealand kakapo is a bird no-one can help but love. About the same colour, size and shape as a very mossy stump, this 'teddy-bear' bird is renowned as the world's fattest parrot. And when survival of the fittest took a turn for the extreme, the kakapo found itself in some pretty deep trouble.

When the first settlers arrived in New Zealand two hundred years ago, kakapos were everywhere. One early explorer described the noise they made as "screeching and yelling like lots of demons." Then again, he also mentioned shaking them like apples out of trees. With the settlers came cats and other invasive predators that rapidly spread across this virgin land, decimating all sorts of birds including the kakapo.

Kakapos have the most immaculate camouflage amidst the lush forest greens, but they had no experience of mammal predators. Let it get close enough, and even the most short-sighted cat will spot a parrot in hiding. Without a 'fight or flight' response, the birds were defenceless. Left alone, a kakapo can live for one hundred years. The flipside of that is that the parrots are extremely slow breeders. They also come to the ground for courtship and nesting, leaving them exquisitely vulnerable to predation.

By 1995, a mere fifty of the birds remained. These were caught and relocated to three tiny predator-free islands off the New Zealand coastline. There began an operation of such complexity as to put to shame the very best spy film plots: operation Save the Kakapo. To start with, every one of the relocated birds received a satellite tag. These cute little backpacks allowed scientists to track each of the individuals' whereabouts. They were also equipped with a remote key that opened each bird's personal feeding station- strictly tailored to that parrot's personalized diet plan to keep it in prime form.

Previously, 40% of kakapo eggs were found to be infertile due to inbreeding. So sperm was collected from the males and flown via drone between islands. Each delivery was matched to the female it was intended for, with an efficient artificial insemination taking place. Every kakapo's genome was sequenced, and matches made strategically to promote the maximum possible genetic diversity. Finally, once a female had laid her clutch, the eggs were taken and incubated, replaced with 3D printed models that moved and made noise in a realistic fashion. The stolen eggs were hatched and hand-raised except for one chick per clutch that was returned to its mom, giving her plenty to keep her hands full.

With all this effort going into reviving the kakapo, it's hardly surprising that their numbers have improved. The population has more than doubled overall, in spite of their painfully slow breeding cycles. Some would consider this elaborate work to be overkill, like the panda nannies in China caring for the species' cubs. I personally find it heart-warming that such great effort should be made for a single species, even if part of the cause is its cuteness.

Incidentally, another New Zealand conservation first is the construction of playgrounds for birds. Another type of parrot, the Kea is benefitting from this new infrastructure intended to keep the pesky birds from wandering onto roads. It might not be as high-tech as the kakapo's contingency plan, but it is another pioneering strategy that is actually pretty darn clever.

There you have it. Enough now with technology. This chapter is about collecting stories of success, of hope and innovation. I'll share a few more with you. Some of the most interesting cases are the tales of animal celebrities.

Two generations back, Elsa the lioness was probably the most famous animal on Earth. Her story began in 1956, when her mother was shot by a game warden in Northern Kenya called George Adamson. The lioness had charged George and his companion, but only later did they realize the reason for this behaviour: she was trying to protect three tiny cubs. One of these was Elsa. George brought back the cubs to his wife Joy and together the couple set about hand-raising them on their home in Meru Reserve. The two oldest were eventually relocated to a European zoo, but the youngest- Elsa- remained on the property on account of her age and gentle disposition. Joy was determined to release Elsa back into the wild and let her lead the life of freedom she might otherwise have had.

It was to be a long and gruelling task to prepare her. Despite her "perfect manners," Elsa had grown up accustomed to humans and had to be taught to fend for herself. Preferring to sleep on a camp bed with Joy's thumb in her mouth was not a good start. The couple took Elsa on safari, letting her develop hunting skills and leaving her for weeks at a time. But all-too-often the inexperienced Elsa would return weak and emaciated, clearly unable to survive out there alone. Joy and George were forced to persevere for many months, but finally, Elsa found her freedom and reared three cubs of her own on the reserve. In doing so, she became the first lion to be successfully released back into the wild. Joy and George continued to meet with her until her death, and Joy's book *Born Free* about Elsa spent a full three months atop the New York Times bestseller list.

In 1966, a film adaption of *Born Free* was released, winning numerous awards and proving a hit at the box office. The actor couple that played George and Joy were so profoundly affected by the experience that they established what was to become the Born Free Foundation. The charity now boasts 100,000 supporters, millions of dollars' budget and a critical role as the Amnesty-equivalent for animals. Some twenty lions were variously cast as Elsa for the production, but throughout the filming the actors refused any protection whatsoever from the animals. They developed a very close relationship, describing it as "agony" when eventually they had to part. Hence the true story of Elsa and its filmic reproduction both touched the world, paving the way for future animal celebs.

Like AC-9, the California condor. Firstly, AC is an unimagi-native abbreviation for Adult Condor. Thus excepting a vague

Star Wars feel, AC-9 wasn't done any favours by his naming. What saw him rise to fame was the fact that he was the last condor to be taken into captivity in one of the most daring last-ditch conservation breeding efforts of all time. But let's rewind a little and take it back to the start.

Condors have graced the clifftops of California and Arizona in the US for the past 15,000 years. With a three-metre wingspan, their silhouette is simultaneously breath-taking and imposing outlined against the sky. Unfortunately, not everyone takes kindly to such creatures, and lead poisoning from hunters ranks chief amongst their threats. By the 1980s, California Condors were in steep decline. A hotly contested debate broke out on whether or not to capture the remaining two dozen birds for captive breeding. This bold move would be practically unheard of, but the evidence was clear that without drastic intervention the species faced imminent risk of extinction.

Slowly, tentatively, condors began to be taken in. At first it was the young birds and eggs from breeding pairs, but when six wild, adult birds went missing within a six-month period and five breeding pairs become one, matters became tense. Four males and one female were all that remained of the wild condor population. AC9, a young male at the time, somehow beat the odds to win over the aged female AC8. The couple produced two chicks together. Too little, too late.

The call was made to take the remaining birds into captivity. AC8 was removed, then two of the males. AC9 watched from a nearby tree as the second-last condor (AC5) was captured. He then managed to elude all efforts to trap him for months thereafter. Finally, on Easter Sunday 1987, he was feeding on a goat carcass laid out as bait when a net was blasted over him from canons hidden in the rocky landscape. Condors were extinct in the wild.

The race began to breed them back up in zoos, with 22 of the birds then held across zoos in San Diego and Los Angeles. AC8 and AC9 were separated to maintain genetic diversity, both of them going on to lead the breeding efforts of the birds. AC8 raised 12 chicks in almost as many years, before becoming infertile due to age. She was released in 2000, the oldest condor then alive, with a personality described as shy but smart. AC9 was by all accounts curious and bold, adapting well to his new surrounds and later acting as an adopted parent for numerous younger birds. He produced 15 chicks in 15 years spent at San Diego Zoo, and was released in 2002. The first bird to be radio-tagged in the -80s when condor monitoring started, his tracker stilled in 2017 with his death. Lest we forget.

The Condor Recovery Program is and was a mammoth venture in the history of US conservation. Costing tens of millions of dollars over several decades, the result is that now approximately 500 condors grace the skies and releases are ongoing as far away as Mexico. The only effort that could dwarf that afforded to condors, in courage, innovation and resources, is that made in the uphill battle to preserve coral reefs. So it is only fitting that we end the chapter with an exploration of this troubled front, and the successes achieved there in the face of terrible odds.

Coral reefs are important. They provide $12 trillion annually to the global economy, with billions going into the livelihoods of local communities through tourism alone. Most are between 5,000 and 10,000 years old, but some stretch back tens of millions of years. They contain more than one-quarter of all ocean life but cover less than 0.1% of its area. In short, they're well worth protecting.

The trouble is, we're not doing a great job at that. Half of all coral reefs have been lost in the past 30 years, and by the time global warming hits 1.5 degrees Celsius, up to 90% are expected to have died. Rising ocean temperatures cause coral polyps to release the tiny algae they live alongside, losing their colour in a process known as bleaching as they slowly starve to death. That's shorthand for 'we need to act.' Urgently, large-scale, cleverly and now. The good news? We're already most of the way there in developing the solutions needed to address this global crisis.

First off the block is the obvious solution: why don't we just collect coral eggs and sperm and 'captive breed' them too? We do. In nature, the success of these species is based on a 'more the merrier' approach. Once a year, close to midnight with a full moon overhead, all the corals in a reef spawn simultaneously. Hundreds of millions of eggs and sperm are released, and other marine creatures go to town gobbling them all up as fast as they can. Only 0.2% survive long enough to become corals, but if you can get out on the reef at the right time and collect the sperm and larvae, then 90-100% can make it through. That's a lot of corals.

All sorts of tools are used to do the collecting, from tents put around the corals to large sweeping nets for divers, floating contraptions and submersibles. Once the corals are large enough to fend for themselves, they're reintroduced. They may be stuck to cheap concrete tetrapod structures and thrown in amongst surviving corals, or better, relocated via underwater drones to safe spots where they can grow. This helps replenish dwindling populations of corals and restock crucial marine habitats, which could just buy us time to save what's left.

Another technique that goes hand in hand is that of micro-fragmentation. As with most brilliant ideas, this came about purely by accident- a scientist in Florida accidentally crushing a coral trying to pick it up out of a tank. For weeks, the fractured coral shards remained unnoticed and undisturbed until the scientist in question happened to pass by the tank once more. Miraculously, they had not only regrown but had done so at a rate more than 25 times faster than usual. A full year's growth has been made in two weeks, and in doing so they had recombined into a single cohesive structure.

This finding was more than enough to widen the eyes of conservationists worldwide, and projects emerged making the most of this strange phenomenon. Specialized saws can remove groups of 1 to 5 polyps from living corals, which are then placed in shallow water to grow before being released back onto the reef as fully-formed corals within months. The challenge with bleaching is that bleaching events are now happening much faster than the reefs can naturally recover from. However, aiding the process with micro-fragmentation, reefs get a much-needed boost in the ongoing race.

In the tiny island of Grenada in the Caribbean, 'biorock' reefs take this to the next level. Biorock, or 'seament' as it's otherwise known, is a sort of artificial reef that starts with a steel frame dropped into the water. A small electrical current is passed through and minerals accumulate from the surrounding seawater due to natural chemical processes. The ingenious thing is: these minerals are precisely what reefs need to grow, and so corals develop rapidly around the frame. In other areas, the corals themselves are 3D printed, with the same size, shape, texture, and chemical makeup as the real things. Very clever.

Other projects work on a different level, mapping reefs for research and educating the public about the dangers they face.

NASA has kitted out a plane with a world-leading remote sensor device that creates detailed 3D renderings of reefs from the air. Google has pioneered a 360 degree virtual reality underwater camera, which is travelling around the 50 most protectable reefs on the planet, capturing them in exquisite detail and uploading the results to Google streetview. Some corals can be saved by breeding them faster than they'll die off, but ultimately the survival of today's reefs depends on a comprehensive understanding of the threats they face and the spatial patterns of surviving corals across the world.

Perhaps most ambitious of all are attempts to create 'supercorals' through so-called assisted evolution. Just as we altered wild plants and animals into the foodstuffs we know today, some scientists are selectively breeding corals for the maximum possible resistance to warming oceans. They also train parent corals to cope with higher temperatures, allowing them to pass on even greater abilities to their offspring. The hope is that these elite corals will form the backbone of future reefs, and with climate-control efforts to back them up, they might just stand a chance.

If this chapter teaches you anything, it should be that there are lots of people who care a hell of a lot about nature. When we put our minds to solving some of the major issues faced by biodiversity, then anything is possible. Great changes come in small packages, and ideas are like corals. A breakthrough innovation can spawn thousands of creative, connected, curious, collaborative projects around the world. And eventually, these may add up to saving it. Why isn't everybody on board, rooting for team nature? We'll see...

8

Let's Eat Granny

Australia is a land of monsters. From the man who was bitten twice on the penis by a deadly redback spider, to jellyfish stings described as 'like being hit by a car.' From octopuses capable of killing 26 grown men in minutes, to the case of an adorable koala mauling a woman's leg- there's plenty to watch out for in this Pandora's box of a continent. Falling prey to one of these nasties, you could be in store for anything from breathing difficulties, minor swelling and muscle spasms to sweating, nausea, headaches, vomiting, internal bleeding, cramps, burning pain, paralysis, dismemberment and cardiac arrest. In short, it's a miracle I'm alive.

One of my all-time heroes is an American entomologist by the name of Justin Schmidt. Beginning in the late 20th century, he set about getting himself stung by dozens of ants, bees and wasps in the name of science. The results form what's known today as the Schmidt Sting Pain Index. Each of the insects is ranked on a scale from 0 to 4, with 0 being imperceptible and 4 promising excruciating pain and lasting agony. The report he wrote to introduce this is full of colourful descriptions of each sting for the curious reader.

As you'd expect, Australia has more than its fair share of these painful stingers. On the lower end is the fire ant at 1.2, with a sting described as "Sharp, sudden, mildly alarming. Like walking across a shag carpet and reaching for the light switch." Moving up the scale, we find the giant bull ant on level 3: "After eight unrelenting hours of drilling into that ingrown toenail, you find the drill wedged into the toe." Ouch! The paper wasp also occupies this level, being equivalent to "spilling a beaker of hydrochloric acid on a paper cut." And finally, on the exclusive upper rung sits the spider wasp, scored at 4 out of 4 and described as like "a running hair dryer has just been dropped into your bubble bath." Poor old Schmidt-he really has done us all a great public service.

Why do I mention this? Two reasons. Firstly, it shows that everything is relative. A fire ant sting may not be pleasant, but compared to that of a spider wasp, it's bliss. Secondly, it shows that we have a strong learnt fear of things we have never experienced. At least, I'm hoping most of you haven't experienced all of these stings. Both of these principles will be important for what is to follow, because this chapter is all about setting straight some of the myths and fears that surround animals. We'll dive deep into conservation's dirty work and the parts of human nature that needs fixing.

Here in Australia, despite the vast array of bites and stings you're prone to receive, there's one nasty offering we miraculously lack: Lyme disease. Caused by the spread of bacteria from ticks, there is thus far no conclusive evidence of the disease being contracted on the continent. It's found across Africa and Asia, Europe and the Americas but not Australia, which is a blessing. Sufferers of the condition, left untreated, can experience searing pains, partial blindness, panic attacks, migraines, depression, delusional behaviour, and heart failure.

I can understand that people would fear ticks for this, yet Aussies are just as tick-wary as any nation's citizens. Why? Part of it must be the fear response that's triggered by the idea of an animal sucking your blood, be it mosquitoes, ticks or leeches. Increasingly, though, there's another factor coming into play. It has to do with another tick-borne condition that manifests itself very differently.

Alpha-gal allergy was first recorded in 2002 in the United States, and five years later in Australia. In simple terms, when certain species of tick bite a human, a small amount of a particular carbohydrate called alpha-gal is released into the bloodstream. The human immune system, misinterpreting this as a potential threat, seeks to eliminate it and so develops defence systems called antibodies by which to do so. The thing is, these antibodies remain in the bloodstream permanently. Any time the affected person eats red meat- which is high in alpha-gal carbohydrate- the antibodies are activated and switch the person's immune system into overdrive. They become allergic to the stuff.

I'm not vegan, nor am I strictly vegetarian (though I have the utmost respect for those who are). Even so, I don't see a red meat allergy as too horrific a condition. The ticks are welcome to do their worst and I'll live with the consequences. That's not the case with most people, and Aussies love their red meat. For many, the idea of alpha-gal allergy is anathema.

And that raises an important question: to 'save the world' and live by a Human Nature worldview, would we all have to become vegetarians? Would we all be forced to inflict this dreaded condition upon ourselves? There's a lot of sense to that fear. Meat products have a high impact on environments, and many inflict some form of cruelty to animals. The answer would appear to be an obvious 'yes.' My answer is 'no.' And there are many reasons why.

First some background. Fears about food supply can be traced back to the great British economist Thomas Malthus two centuries back. In his 1798 book *An Essay on the Principle of Population*, he set out all the horrible ways in which communities that expand too fast collapse due to nature's checks and balances. The most prominent of his theories, for which he's known today, is the Malthusian trap. To understand this, imagine a graph. One line shows our global food production, which increases at a (roughly) linear rate. A second line shows our population, which as he knew, is growing exponentially and so looks like one side of a steep skateboard ramp. At some point, these lines will cross and the result is described alternately as a 'crunch,' 'crisis' or 'catastrophe'- none of them particularly inviting terms. When we have more mouths to feed than food with which to feed them, society implodes.

What Malthus failed to take into account is human ingenuity. In the 1950s and -60s, the Green Revolution took place. Novel farming techniques came to light with mass fertilizer usage, new crop varieties, and improved irrigation techniques. The result was a boom in agriculture across the world. In Asia, yields of wheat and rice doubled, reducing prices, cutting poverty by half, and seeing the entire growing population receive one-third more calories per head than they previously had. This Green Revolution came at a time when global food supply was coming to a head, and its importance cannot be overstated. One of its initiators won a Nobel Peace Prize, credited with saving 1 billion people from starvation.

Agriculture's remarkable reimagining allowed our population to continue its steady climb. With the current rate of population growth, we're going to need another revolution to keep us all fed by 2050, but the technology for this is pretty much already here. Leading the field are the brilliantly controversial GM crops made by huge companies like Monsanto. These are plants designed for maximum efficiency, their genetics edited for disease resistance, resilience against natural disasters, and more grain production per unit land. They're controversial because they monopolize the industry, and because people see them as downright unnatural. People fear these so-called 'frankenfoods' because they're new and sciencey (technical term), but our food hasn't been natural for thousands of years, given the selective breeding farmers use to maximize their produce.

Grains and meats aren't equivalent, but they're more closely linked than you might think. Ever wondered what cows eat? Because I guarantee you it's very rarely grass. In ecosystems, there are four levels on any food chain, and the higher up you go the more energy is required to sustain the lifeform. Plants sit at the bottom, which makes them efficient as a food source, whereas cattle, sheep and pigs are more demanding on resources at one step up.

BUT if we use more nifty, creative means of supporting these animals, we're able to offset some of those environmental impacts. Feeding seaweed to cows drastically reduces their methane emissions, and these animals too can be genetically 'improved.' They still take up land space that could be used for planting crops, but as the famous impossible burger shows, the future of meat might not involve cows at all. Patties can now be made at competitive prices to taste like beef without any trace of meat, wiping away any questions of animal cruelty. Perhaps in the future, we'll all be vegetarians without knowing it!

And there are two considerations that seal the matter for me. Firstly, a vegetarian diet is unnatural. It's no surprise that some people struggle with it, as Homo sapiens evolved as an omnivore. We're built to live off both plants and meat, even if the latter was in lower quantities than most people consume today.

Secondly, following a vegetarian regime can be highly counterproductive in minimizing your environmental impact. Where do dairy products come from? Cows. So eating more dairy products like cheese and yoghurt to make up for losing meat proteins increases the diet's negative impact instead of the reverse. Greenhouse gas emissions from your food are largely contributed by sourcing overseas ingredients, so eating locally-farmed meat is far better than importing foreign veggies. The compromise I'm in favour of is the planetary diet, with a few serves of meat per week and plenty of legumes. Although I wouldn't say no to one of NASA's 3D-printed pizzas...

All this fretting brings me to another very important point. Populations are not going to keep on rising. Seeing the spike in any population graph, a blindfolded four-year-old could continue the line. What does take a bit more thought is to match this with the declining birth rates globally and to realize that our population is already starting to level out. Once 2100 comes around, that line on the population graph will have flattened and the total human population will rest at about 10 or 11 billion. For me, that's very reassuring. It means if we can find a good way to feed and clothe and house all those people, we can set about making some big improvements to our environment that'll last. Just one more fear we needn't be so worried about, and in fact, a very good reason to look ahead and find the solutions now to build into our plans.

If you took all the mammals on this Earth and put them on a scale, 60% of the weight would come from our livestock and 36% from us humans. That leaves only 4% for the thousands of species of wild animals, an altogether paltry amount. Amongst birds, it's 70% for chickens and 30% for everything else- again an unnerving comparison that leaves much to be desired. So, whilst the Human Nature worldview wouldn't mean everyone going vegan, whilst our population is plateauing and whilst my faith in technology is sound, I do hope in a few hundred years we'll have flipped the scales and left the wild animals at a share of more than half.

Enough of that. This chapter is about the forces holding us back, and whilst food is a troublesome issue, it's hardly unsolvable. One of the queries I find hardest to address is what I call the Mosquito Conundrum. And it goes waaaay back.

I doubt you need to be told that mosquitoes are bloody nuisances in every sense. There's a wild claim that they've killed half of the humans that have ever lived, although it's hard to know for sure. What's certain is that the malaria parasite they carry is the most deadly disease in human history. It has killed at least 54 billion people to date and continues to account for 1% of deaths annually. Half of all people are at risk of catching it. Four US presidents have been infected with malaria, it killed Tutankhamun and Alexander the Great, stopped the armies of Attila the Hun and halted Genghis Khan's world conquest. Quite the bucket list for a time-travelling megalomaniac, and all caused by a teeny little microbe shorter than the thickness of a piece of paper.

The question is: should you swat and kill a mosquito? The answer is no, and you shouldn't need to at all. Besides the maddening buzzing noise they make, it's the blood-sucking part of mosquitoes that makes them our enemies. Over the past century, we've coated them and us in the toxic insecticide DDT that kills most forms of life and effectively kickstarted the environmental movement from campaigns against its usage. Nowadays, that's not an option. But modern deterrants are much better, and are even being built into clothing and wearables. It's a win-win situation because you stay bite-free and the mosquito avoids a swatting. Harmonious coexistence doesn't mean letting animals run your house (though I've been known to do that on occasion). It means accommodating our own needs and wants with those of other lifeforms and finding compromises.

There are more creative strategies. These range from the less practicable- breathing less, so the mosquitoes aren't drawn to the carbon dioxide of your exhalations- to the downright ingenious- putting out old clothes to attract the jumping spiders that prey on malaria-carrying mosquitoes. Drinking less alcohol reduces your chances of being targeted, as does wearing floral perfume.

The most basic understanding of ecology allows us to devise safe and effective alternatives. For instance, we can create garden habitats that encourage dragonflies to breed, as each animal picks off hundreds of mozzies each day on the wing. Microsoft are openly recruiting mozzies as allies in Project Premonition: taking certain species and looking at the blood samples they've ingested for signs of animal diseases that might spread to humans. 75% of all new infectious diseases spread to humans from animals, and the simplest solution is to keep our ecosystems healthy to prevent their breaking out. COVID can teach us a lot. I should also mention that it's only female mosquitoes that bite you, and they need your blood to raise their young. Would you really be so heartless as to deny a pregnant lady her breakfast?

Everyone has their own limits as to how far they'll go to accommodate other animals, and so long as you're considering it that's fine. If you're still sitting in the too-hard-won't-bother camp, let me tell you about the Jain monks to put matters in perspective. Remember the first lesson we learnt this chapter: everything is relative.

Jainism is a small Indian religion, closely related to Hindu and Buddhist teachings. Its monks have to swear to abide by five 'Great Vows,' the first of which is "To injure no living being by action or thought." This rule called "Ahimsa" is followed in the strictest sense, bearing direct and far-reaching consequences for their lives.

Firstly, Jain monks are vegetarian, often vegan. They also refuse root vegetables on account of the tiny organisms harmed when the veggies are pulled from the Earth. They are taught not to leave their houses during heavy rains or monsoon season for fear of upsetting microbes in the puddles, and sweep the streets when they do go out to ensure they do not step on any creatures. Some choose not to go out at night when they are less able to see the animals they may step on, and masks are always worn to keep them from inhaling tiny creatures. Now, if the Jain monks can build a thriving religion off the back of such harsh principles, I think you can manage a few simple courtesies to animals. Don't you?

When "can't be bothered" doesn't stand up to scrutiny, the fallback excuse used for not embracing conservation is that doing so disadvantages humans. It's apparently a "them" and "us" scenario, where helping out animals in need stalls human development. The opposite is true. Biodiversity loss impacts progress towards 80% of the Sustainable Development Goals, and ecosystem services provide 40% of the world's economy. Countries with strong conservation measures also have low wealth inequality. That's no coincidence. As clichéd as it sounds, we rely on nature for everything from the food we eat to the air we breathe to the ground beneath our feet. Let me tell you about Madagascar's lemurs.

I've always wanted to visit Madagascar. It's a land of exotics, left to evolve on a tangent over 88 million years of isolation. Coral reefs and tropical forests are the two most biodiverse ecosystems we know, and here they are both in close proximity. But the face of the island's biodiversity is its lemurs. The word 'lemur' comes from the Latin word for 'ghost,' their queer appearance having earned these enigmatic creatures their name. Sadly it's taking on a new significance, as lemur species disappear. Of 103 lemur species on the island, 23 are critically endangered, 52 are endangered, and 19 are classed as vulnerable. New species are constantly being added to the list and the death toll keeps on rising. They are the most threatened group of mammals on Earth.

How do you save monkeys? You help the apes– humans, that is. The leading cause of decline in these species is habitat loss. 80% of Madagascar's forests have been destroyed. So local farmers are being taught to raise fish instead of crops, given alternate employment through ecotourism, and provided with more efficient stoves that don't require wood fuel. If enough money goes into 'saving humans' and assisting impoverished communities, the drivers of deforestation on the island vanish. This is a growing branch of modern conservation, and it's firmly focused on helping people.

We're not compromising human lives to help out animals; rather we're improving them. And in doing so, we're probably also increasing our own survival chances. 75,000 years ago, an almighty eruption took place in modern-day Sumatra. The supervolcano that blew released a volume of magma equivalent to 1 million times the volume of the Great Pyramid of Giza. It was enough ash to cover Southern Asia in a layer 15cm thick. The planet cooled rapidly due to the sunlight the ash blocked out, and average temperatures fell by anywhere from 2 to 15 degrees Celsius for years after the event. This volcanic winter was accompanied by major ecosystem collapse. In short, it was big: 100 times greater an eruption than anything we humans have experienced in recent history.

As you might expect, early humans faced some problems of their own at the time. What with the chilly temperatures and scarcity of game, our ancestors' population shrunk to as few as 3,000 people. Drastic decreases in population led to low genetic diversity, creating a major stumbling block in our evolution. Just after humans left Africa, the same thing happened. The population shrunk into a bottleneck, and every person alive today can trace their lineage to a single mother around this time. Research has shown that a group of 55 chimpanzees would have more genetic diversity between them than all 8 billion of us modern humans. That's how close we came to going kaput. And it's not unthinkable that we might face another similar situation in the near future, where the resilience of our ecosystems and communities is tested.

What if a very large eruption happened nowadays? As it happens, we do have an active supervolcano hidden under America's most famous national park: Yellowstone. If it erupted now (and it could) there'd be some large, immediate effects. Shedloads of molten rocks and ash would be flung into the air, covering much of North America. Any tourists there at the time would die horribly, but what about the rest of us?

Thankfully, we'd deal with it a lot better than we did with its Sumatran cousin. The global economy would take a blow, but recover after several years, and the world's food supply would also shift accordingly. Some of us would go hungry, but most people, on the whole, would be okay. The reason for this is our development century by century, learning resilience the hard way and working with nature. We've built up systems that offset and distribute our environmental impacts, improving sustainability. Conservation strengthens this and thus may end up saving our skins.

We're making our way rapidly through anti-environmental excuses. Let's take aim at a more unusual one. Could animals take our jobs? Right now, it's mostly robots we're worried about, that are sucking up manual roles in the labour force and putting some people out of work. Animals can do many things better than robots and people, so the question isn't quite as outlandish as it sounds. That said, I'm pretty confident in saying that at this stage, your company's board isn't going to be replaced by a panel of chimpanzees. Would animals be able to enter the human workforce?

The first role I reckon they might interview for is as inventors. I've already mentioned back in chapter 4 how humans took the better part of their ideas from animals, so I'll skip the sermon. Only consider the tiny gall wasp. It's a species that inject its eggs into trees in such a way that the tree grows a protective casing- a 'gall'- around them. Then consider the thousands upon thousands of galls that were harvested to produce ink for 1400 years of recorded human history. That's a lot of royalties. Gall wasps were responsible for the oldest surviving Bible, the complete works of Shakespeare, the first dictionary- you name it. Sadly, gall wasps lack in legal representation, and their invention never earned them a cent.

But what if our society recognized animals' contributions? It's a radical, yet not unimaginable line. Humanners, the idea that animals deserve treatment as fellow people, would suggest that we pay them such respect. We've learnt water filtration and electricity generation from nature, and the food we eat is altered using nature's methods from long-lost plants. We also get most modern medicines from nature, though it gets precious little in the way of thanks. One existing project that's working to turn the tables is the Lion's Share Fund. Big-name corporates like Mars and The Economist have signed up to this collective, agreeing to share a percentage of their advertising bill for conservation whenever animals feature in their promotions. Neat.

Legally speaking, there isn't anything major that prevents other animals from taking up places in human society. Chim-panzees could theoretically testify in court and are increasingly being recognized as legal persons. Note that a 'person' is different from a 'human,' however the status affords an animal most of the basic human liberties. It has been suggested that chimps actually belong to the genus *Homo* alongside ourselves. That would mean that humans a third kind of chimpanzee (two chimp species exist in the wild). Searching job sites with the query of 'chimp job' reveals some amusing finds: website maker, ticket salesperson, social media manager and more all apparently fit the search. And I for one think it would be charming to be met by a chimpanzee over the counter at a cinema. They are stronger by weight than we are, so would do admirably in heavy-duty construction. And in a chimp-human olympics, I'd place money on the chimps coming away with at least a few medals.

I mention these hypotheticals to reflect just how much our attitudes towards other animals have changed in the past few decades. From denying they could make tools for most of the 20th century to acknowledging their sentience, admitting we may share a genus and that they possess abilities we can't compare with.... that's a big leap. Looking forward, it's hard to know what's next. I don't think it's necessary to integrate apes into our workforce, nor do I think anyone ought to concern themselves with the prospect. I do believe that they might benefit from our human progress in other ways, via technology and new communication avenues.

There are animals beyond our closest relatives which have proven themselves adept at certain tasks. Paul the octopus was kept in a German aquarium during the 2008 football world cup. As a gag, his keepers designed a betting game, whereby Paul would be given a choice of food containers representing the two teams in an upcoming match. The flag on the food container he chose to eat from would be taken as the country he thought would win. It turned out that Paul had a knack for predicting results. Had the wages been real and not hypothetical, a $100 starting bet could have earned him $200,000 by the end of the tournament. Impressive!

Pigeons have been shown as proficient art critics: they're able to distinguish between good and bad art, pick out impressionist works from cubists, tell a Picasso from a Monet, a Braque from a Cezanne, a Renoir from a Matisse. That's more than I could do. In medicine, their perceptiveness allows them to match experienced doctors in giving certain diagnoses. Provided with microscope images of human breast tissue, pigeons are able to correctly state whether a malignant cancer is present. Individually, their accuracy is 85%, but teamed up in groups of four, they reach a stellar score of 99% correct diagnoses.

And last but not least, there's Hans the horse, who proved himself quite the polymath. Taught by a high-school teacher a century back, it appeared that Hans could perform complex sums, spell human names, read written questions, work with colours, calendars and playing cards, and distinguish between the music of different composers. He became a minor celebrity in his time, though the scientific community was incredulous. Clever Hans was subjected to 18 months of testing by scientists who travelled from across Germany to see him. None of them could find anything that aroused suspicion in his methods.

Eventually, one tester decided to remove all people from Hans' surrounds. The horse faltered and began to lose his abilities. Unbeknownst to his trainer and the scientists that had come to test him, Hans had been reading the facial expressions of his questioner for the correct answer. To respond, he was required to tap out a specific sequence with his foot, like a multiple choice sheet in morse code. As he reached the correct number of taps, the face of the questioner would relax imperceptibly. Hans would then stop, amazing his audience. Brilliant!

If you're not concerned about having your sketches poo-pooed by a pigeon or having your mind read by a horse, then you might be wondering what the catch is. Surely everyone can't win with this Human Nature thing?

As I'm writing this, we're in lockdown due to the coronavirus outbreak. Toilet paper is in short supply- hoarded by preppers desperate not to go without. A fine moment for the history books. Toilet paper accounts for the loss of 27,000 trees per day, which is a big number on the face of it. You know what's coming. Would a Human Nature worldview mean that we have to bite the bullet and find a new alternative to wipe our read end with?

You would not believe how often I've been asked this question, and I can't help but find it entertaining. Let me endeavour to respond. Firstly there are 3 trillion trees on Earth right now. That's one-half of what there was before we humans came along, but still 400 for every person on this planet. Right now we're losing about 15 billion trees per year, which is too many. But relative to the 3 trillion total, it's not as many as it might sound. Forest loss itself has halved in the past 30 years. All of which is to say that toilet paper has very little effect on tree-loss, accounting for less than 0.001% of deforestation. It's your printing paper hoards and IKEA bookcases you should be worried about.

Let's flip the narrative. Instead of going without toilet paper, what if every one of us planted one tree every other day for a year. That would mean adding more than a trillion trees to the globe. By the time the trees matured, we would have cancelled out an entire decade's worth of carbon emissions. New swathes of habitat would have been created or restored, and so biodiversity would benefit as well.

Are you excited? I am! It feels like Christmas all over again with so many trees. And dare I say it's starting to seem like there really might be a silver lining to this whole idea. One concern remains, and that's the elephant in the lobby. How much space exactly does nature need?

Half the Earth, and I'll tell you why that's fair. Say there are 1 trillion species on this planet. That means 0.0000000001% of life is us humans. It could be much less if you count individual animals, but that's a small enough number for me to work with, so let's stick with it. Right now, about 50% of Earth's land surface is said to have been 'altered' by human activities. That's not fair, is it? Even if you accept we're a lousy stinkin' good-for-nothing species (and I repeat that phrase every day in the shower) I think we can do better.

One-quarter of the Earth's land area houses 80% of its bio-diversity. A lot of that land we haven't mucked up completely yet, which is very good news. Edward Wilson is one of the most famous biologists alive, and has done us all a favour by running the numbers. If we continue with business as usual, that means we can expect 90% habitat loss and only half of all species to survive. If, however, we set aside 50% of our lands and oceans for the benefit of biodiversity, we raise that figure to 85% of species that are able to get by. Bearing in mind the way we're headed, with extinctions happening one thousand times faster than normal, saving that 85% would be a major achievement in itself.

This isn't going to be a random half of the Earth selected: it's a strategy that picks the most valuable half for nature, giving it the maximum possible impact. These places are all around the world and include many of the places we've travelled to in the preceding chapters: Madagascar, Gorongosa national park, the Galapagos and more. Others include California's redwood forests, the Amazon basin, Borneo, the Serengeti, and Central

America's cloud forests as key biodiversity hotspots. In many cases, we wouldn't be establishing new protected areas but expanding upon and improving the management of existing ones. Other times, new ground would have to be broken by asking governments to set aside ecologically important areas for conservation. Is that so unreasonable?

If we were to do this, we wouldn't be forsaking our own comfort of living, we wouldn't be restricted in our freedom of movement, and we wouldn't be displacing Indigenous peoples or rural communities. It's a rare situation in policy, but this really is a win-win/ lose-lose matter. I won't ask you which option you'd prefer. Simply thinking through choices, looking beyond the immediate future and choosing the people with the right ideas to lead us is what we need. Remember the butterfly effect from chapter 1? Small actions make a big difference. Let me tell you a story.

Eugene Schieffelin was a respectable man. Born in 1800s New York, he was the seventh son of a city lawyer, owner of a large pharmaceutical enterprise, and husband to Eleanor Roosevelt's great-aunt. If you looked his name up today, though, you would find it buried under certain other matters. Because the actions of this American entrepreneur left a mark on his country just as fresh today as the day it was made.

It started in his 40s when we founded and became chairman of the American Acclimatization Society. Such groups were relatively common in those days. English colonists felt homesick and thus sought to add touches of their homeland to the surrounding environment. That was Schieffelin's intent when, on 6th March 1890, he visited New York's central park and released 60 starlings there which he had imported expressly for that purpose. More than a century on, we can appreciate in hindsight the monumental mistake that was. The invasive starlings thrived, now numbering 200 million birds spread across almost every US state. They cause $800 million USD of damage every year, were responsible for an airliner plane crash, outcompete native birds for nesting space, cause fires, devour crops, spread disease– you get the picture.

Schieffelin couldn't have made a worse choice. And it didn't stop there. An admirer of Shakespeare, Schieffelin had decided to introduce to America all of the birds mentioned by the Bard in his works. As any scholar can tell you, starlings bear a single mention in the play *Henry IV*, which was enough to pay their ticket to New England. Other species he introduced like the skylark, the nightingale, song thrush and chaffinch thankfully didn't take hold, but the sparrows he introduced formed a plague in themselves and now number 540 million.

If you thought I was going to tell a happy story, then I'm sorry. That was last chapter. Sadly our relationship with the natural world hasn't always been that great, and even misguided appreciation can be a detriment. This story I share as a cautionary tale, because even as we accept the need to intervene to save nature, we need to take care not to make the situation worse. Nowadays, legislation prevents such foolish acts, but illegal trade in wildlife still takes place under the radar. It's a tough balancing act to keep, however, there's one thing which is far worse than ignorance, and that's fear.

Think back to the start of the chapter. I was setting out at length the danger of the land Down Under– the Australian outdoors. Thankfully, I'm not dead (yet), and part of the reason lies in the fact that danger is relative. That's lesson 1 again. Here in Australia, more people die from "contact with the hot water tap" than from venomous spiders. Shark attacks are infinitely less likely than injury or even death from falling out of bed, yet which are you more afraid of? True, there are some unpleasant ways to go in nature, but far worse deaths have inflicted by our fellow humans. We don't fear for our lives every time we walk down the street.

More than anything, the modern fear of nature is a fear of the unknown. If you had a pet shark (and don't we all wish we did?) then chances are you'd be slightly more comfortable meeting its cousins in the wild. The 2007 *Guinness Book of World Records* lists the Cassowary as the world's most dangerous bird, but this shy and solitary creature is far less likely to kill you than a crow to the windscreen whilst driving. I read recently that wolves kept in sanctuaries love cream cheese, and I found myself laughing because that image is entirely at odds with our stereotype of the animals.

One story that never ceases to amaze me is that of two girls- Amala and Kamala- as recorded by an Indian missionary in the early 1900s. Abandoned by their mothers at a young age, these girls were adopted by a wolf mother and raised as her own until they were found over a year later by the missionary. He took them in himself, but these 'feral children' had not been raised by human hands, and so acted in the manner of wolves. They refused clothes and cooked food, walked on all fours, had superb night vision, and howled at the moon. Neither of them spoke, and whilst Amala died shortly after from a kidney infection, the older girl Kamala lived for many more years before passing away from tuberculosis.

Whether this story is true or not is hard to tell, yet the likes of it have been told across every culture since the start of recorded history. I don't doubt that such things have happened, and these extraordinary occurrences are a tiny fraction of the well of tales told about animals selflessly helping people. Be it dolphins rescuing surfers from shark attacks or elephants saving weak swimmers from drowning, all these tales share a similarity in that animals are cast as the 'good guys.'

Why don't we keep it like that? In word association, 'shark' invariably matches with 'attack,' making them the typical animal villains. Six people are killed globally every year by sharks on average. That's not zero, but regardless our response to the threat is disproportionate. Yes, sharks have teeth. Not big ones, but lots of them. And yes, a few species of sharks can amputate human limbs. BUT that's a fleetingly rare occurrence, and wouldn't happen at all if people were more sensible.

In Australia, elaborate and entirely ineffectual schemes are used to 'remove' this imagined threat. There's a fine line between policy and paranoia, and I rather fear these measures fall on the wrong side of matters. Millions of dollars annually are spent maintaining shark nets off the coastline. These clumsy underwater rope ladders certainly don't stop sharks from passing and certainly do kill thousands of other marine creatures in the process.

The state government in Western Australia describes their policy as to "track, catch and, if necessary, destroy sharks" in a form of mass culling. That about sums it up. Culling is controversial at the best of times (rightfully so). If an invasive species is ruining local ecosystems, then in extremis a loss of life might be justified. But given that sharks kill maybe one swimmer every few years in Australia, this animal genocide is madness.

Australia is the only country I know of that's crazy enough to continue such regimes. New Zealand removed their shark nets long ago, and Brazil rejected the option outright. South Africa has a simpler and more effective shark spotter programme, with a keen pair of eyes scanning from a nearby hill for sharks approaching its beaches. With improvements in drones and sonar buoys, this entire system could easily be automated. It's not rocket science- it's real science, a.k.a common sense. The single factor that beats science every time is fear, and it holds back conservation much the same.

We fear large things with big teeth, and we fear the unfamiliar. But we also fear the tiniest things and the everyday world around us. We've learnt a new range of fears for our 21st-century existence: chemophobia for the toxic substances we create, acrophobia for the skyscrapers and planes we've built, and germophobia for the microbes we can't seem to ever get rid of even with all this advancement.

There's a substance called DHMO: dihydrogen monoxide. As protesters against its use inform: "Its basis is the highly reactive hydroxyl radical, a species shown to mutate DNA, denature proteins, disrupt cell membranes, and chemically alter critical neurotransmitters. The atomic components of DHMO are found in a number of caustic, explosive and poisonous compounds such as Sulfuric Acid, Nitroglycerine and Ethyl Alcohol." Accidental inhalation of DHMO can cause death whilst prolonged exposure typically results in severe tissue damage. In gaseous form, it causes severe burns and forms a major component of acid rain. It corrodes metals and contributes to soil erosion, initiates violent weather patterns, is used by radical political groups during rallies and sustains all biological weapons.

In case you haven't heard of this substance before, it's otherwise known as water. Dihydrogen monoxide is its standardized scientific name. The DHMO campaign was a joke concocted by

Tom Way and several other scientists. It started with a survey of the American public that found the vast majority of them in favour of banning the liquid. Nowadays, DHMO lives on as a spoof website and through occasional media mentions. The secret lies in the wording. Water takes lives, but it's also essential to sustain it. Which is important to recognize when we look at the world of the very small: the microbes we so despise.

Keeping with the theme of nasty deaths, say you mushed up my body into one big soup of DNA. Floating around in it, you would find ten times as many microbe genes as you would my own. Most of my cells aren't 'my cells' at all, so to speak, but those of tiny microorganisms that help me to survive. In chapter 1 we touched on what would happen if these creatures disappeared, and to give you a brief refresher it isn't pretty. Complete societal collapse would be expected within a year because 99% of bacteria are 'good guys' that do useful chores around our body.

It's the 1% of bacteria known to cause human disease that are the problem. Very few people would argue with me that it's fair game to wipe these out, but in doing so we need to follow a few basic ground rules. Rule 1: always work for the greater good. That includes both animals and people, because it doesn't pay to wipe out a disease and then find out you've collapsed an ecosystem along with it. Rule 2: work with science, not fear. Rule 3: remember to celebrate!

In America, the white-footed mouse is widely responsible for spreading Lyme disease, which ticks pick up from the infected rodents. Though not often life-threatening, this condition is intensely unpleasant for the hundreds of thousands of people infected by it. Head back to the start of the chapter if you've forgotten the list of symptoms. Having investigated Lyme disease transmission in detail, scientists now suggest that we could greatly limit or entirely eradicate the disease in the USA by tweaking the white-footed mouse's genome. The mouse benefits, ticks benefit, people benefit and the disease loses. My maths isn't great, but I still reckon that's a win overall. Similarly, a gene drive might be sufficient to wipe out chytrid fungus from chapter 3 and save our amphibians globally. In both cases, it's our fault entirely that the disease spread was spread in the first place. So the onus is on us to fix it.

As this chapter draws to a close, I'd like to address the mother of all myths, the dictator of all deceptions, the feudal lord of falsehoods. It goes something like this: "Humans are too small to ruin our planet. Nature changes, and has always changed." The statement is a favourite with climate-change deniers, but can be found throughout anti-environmental rhetoric in watered-down forms. Let me explain.

In the past 500 million years, the Earth has seen five mass extinction events during which 75-90% of all species went extinct. Now, we are rapidly entering into a sixth mass extinction event caused by human activity. This one is happening much much faster than its predecessors. Likewise, the global climate has been through major fluctuations, with five significant ice ages known to science. As the Earth came out of these, it naturally warmed by several degrees. The difference is that it happened over millions of years and that we're currently due for an ice age, not a warm spell. Oops.

As for change, that's true enough. Evolution can be slow, but when the pressure on a species to adapt is sufficient, it can also take place in a few generations. During the 2nd World War, mosquitoes followed civilians hunkering down in the underground systems. There, separated from the world above, they evolved into an entirely separate species within a few short years. When it comes to ourselves, studies show that human evolution recently has accelerated to a one-hundred-times faster rate. Traits such as blue eyes and lactose tolerance have emerged only in the past 10,000 years since we settled down and started farming. Usually, a mammal species will survive for a million years on average, versus marine invertebrates which stick around for around 11 million. In either case, nothing in nature is permanent, and humans didn't invent extinction. We just helped it along.

On which cheerful note, I'll bring this chapter to a close. We've seen that everything is relative, that fear trumps science, that animals are great and humans less so, and that ultimately humans are a cog out of place in nature's ebb and flow. To keep with the rhetoric of conservation as a battle, we need to fight smart. We must mass the infantry, revise our tactics, and win the war. That's what the next chapter is for. It is an action plan, a guide for reformation. One last hurrah...

9

Worth Sharing

Eight years ago, I almost lost my life. Camped with my family in the North of Botswana, I had gone for a nighttime trip to the toilets. The bathroom block was on the far side of the campground, and I was walking back in the dark. The night was still, the bush crickets chirping and thousands of stars looked on overhead. I reached the edge of the fire circle, with its flickering light and dancing embers. On a whim, I turned around. A dark shape lay at the edge of my shadow.

"I think there's something behind me," I said. My brother turned around. There, lit up by his spotlight's pale beam, was a young leopard- crouched, stalking, low to the ground. It stood there, frozen, as did I. Our eyes locked. Seconds ticked by, then the contact was broken and the leopard padded off. I made a beeline for my tent, where the leopard would not follow. But the animal stayed by the campsite all night, circling the fire.

The following morning, there was a post up on a major 4WD forum describing the encounter and warning fellow campers of the like. Immediately, the comments started rolled in, piling up into a thread some four pages long. There was a heated discussion on what might be done with the problem leopard. There were anecdotes from others who'd come across leopards up close. And, amidst it all, there were cutting remarks on what should be done about the problem child (me)!

To clarify, my actions were not unduly brash. In the normal state of things, a leopard would almost never stalk a human-even a child such as myself. Most likely, it had been fed by prior campers, and so grown to associate humans with food. Through no fault of its own, it had become a threat to people. Relocation would require darting it with a tranquilizer- a costly and dangerous process for both the people tasked with the job and for the leopard itself. Besides, anywhere the problem leopard were moved to would likely be another leopard's territory. Shooting it was the authorities' easy option.

This animal failed to take my life and paid the ultimate price. It's a strange sort of debt I feel that I will never repay. No single person or animal can shoulder the blame, yet the tragedy took place all the same. I see it as a symptom of our current relationship with nature- a relationship I have devoted my life to changing. We are the product of our environment, but we also shape it.

In the previous eight chapters, we have taken a joyride through the abundance of life and the shifting sands of where humans have fitted themselves into this picture. This final chapter is where I'll weave these separate strands together, and round out the greater narrative each chapter has been building towards.

Our trip started in the distant past, charting the history of conservation and of pigeons through the coevolution of people and animals. We touched on the 'Scala naturae', the Great

Chain of Being, with the endlessly humble *Homo sapiens* at the top and the damned and the damn hard (i.e rocks) at the bottom. Well, I propose a new Greater Chain of Being Human - a version 2.0 of Aristotle's teachings.

At the bottom of the chain, we have those who exploit the natural world. Here we have the poaching syndicates, animal circuses, illegal loggers, exotic pet traders, and all manner of others. This is the lowest form of humanity, reserved for the broke, the corrupt, the desperate, and emotionless people of this world. One rung up, we have those who show indifference towards nature. These are the restauranteurs who choose fake plants over real ones, the commuters squashing city ants beneath their feet, and the drivers too lazy to swerve for a hedgehog on the road. Too many people sit in this class. It's a privilege our cushioned lives have brought because our connection with nature from behind our urban walls seems somehow reduced. You're reading this book though, so I'll bet you sit on a higher spot. If not, you will soon.

The third rung belongs to those who actively appreciate nature. They might keep a bird-bath in their garden, feed the ducks at local parks, go out tree-planting on occasion, or take long strolls through national parks. This is a growing crowd, and an active one at that, made up of the most wonderful people. So there remains only the fourth and final rung to discuss. You know what this is, having read tens of thousands of words about it. Human Nature. To be here, you've accepted the worldview, embraced Humanners, and made a habit of showing nature gratitude. Amongst much else, I hope.

When we made it at last to chapter 2, the theme became one of human ignorance towards nature's basic facts. It showed our tendency to dramatize bad situations into nightmarish ones and to place the blame squarely on others' shoulders for our shared problems. We saw how rapidly our knowledge of other animals changes- science in this realm is far from static- which brings hope that we might soon understand the meaning of life. Well... almost.

In chapter 3, we swiftly addressed the challenges facing modern conservation, and like Sir David Attenborough's bush-babies, we clung to improving interspecies relations as a way forward. In chapter 4, we saw the genius of animals, learnt how people plagiarized nature's inventions, and then progressed in chapter 5 onto how exactly we pin down 'humanity.' We picked up an arkfull of querky species in chapter 6 and followed their stories to see what they might be able to teach us. Finally, chapter 7 offered up a glimpse of hope through community, science, and technology colliding. And chapter 8 dampened the flame with a review of our doubts and fears. All of which brings us to the here and now. Where eight chapters become one.

I'll give you my password: I PLEAD 4 CHANGE. That's all you need to unlock Human Nature, and to remember 75,000 words of this book's wisdom. It's not my favourite of acronyms, but if we think of ourselves being trapped in The Matrix of our status quo, this is how we wake up. Ready?

I is for intelligence, because we can't go on claiming we're the superior race. That's what we did 500 years ago when European empires spanned the globe and trounced on Indigenous land rights. The arguments we're using today are much the same as were used then: animals are 'uncultured', 'unintelligent', 'primitive', and so on. There is nothing that makes humans intrinsically- morally or genetically- superior to animals, just as there is nothing that made Europeans better than the

Indigenous tribes they met and enslaved. Most people would agree that colonial behaviour during those times was at rock bottom for human etiquette. The good news is, we can learn from that.

Avoiding the usual human bias, the definition I would offer for intelligence is "the ability to sustain a happy, prosperous existence for oneself and one's kind." If you think that's unfair, I could just as easily subvert the old human tactic of defining brilliance as ourselves and say that intelligence is one's similarity to an amoeba. That would be just as fair as our own previous measure.

But we're trying to improve upon it, and all species are striving for a "prosperous existence" in some form. The happiness clause allows us still to appreciate high-performing social animals, whilst encouraging us to widen our perception of emotion in other lifeforms. Mind you, humans wouldn't be bottom of the ladder with our revised form of intelligence. It would simply give us something more to aim for. Having to 'sustain' intelligence is our major challenge. That fits.

Stegosauruses are famous for having brains the size of walnuts, which is both wrong and misleading. Firstly, their brains were closer to the size of a lime. Secondly, these chunky dinos had intelligence without needing brainpower. Each had a thagomizer (spiky tail club) to disembowel anything that attacked it. And with a weapon so fiercely named, it would take a pretty brave tyrannosaur to mess with one of them. The stegosaurus species lasted for 5 million years, marking it down as extremely successful. Compare that to the paltry 0.2 million years that *Homo sapiens* has managed, and we should be venerating these beasts for their longevity. Never confuse brainpower with intelligence.

P stands for policy, which is where we start to prove our newfound smarts. Setting aside our love for cute and cuddly creatures, we need to find better ways of prioritizing so that each conservation win counts for more. Umbrella species like koalas allow resources to go further because protecting them helps a whole suite of other animals. They're like nature's social butterflies, and losing them really spoils the party. Plus they're cute and kinda cuddly as well, so that's a score on both counts!

On the other end of the spectrum, there are cases of special animals that are so unlike any others we know that they merit extra effort to save them in and of themselves. The Chinese giant salamander counts on this list, as does the Ganges river dolphin, the black rhinoceros, and the Bactrian camel. These are mightily endangered, but also evolutionarily distinct- in a word, 'special.' The species' survival plans invariably start with better monitoring, required to avoid the embarrassment of them slipping away under our noses. What's equally important to keep in mind that we've described less than 20% of all animal species. We can certainly pick favourites from those we know of, but let's not do so at the expense of those yet to be discovered.

Protected areas also count under policy (and begin with 'p'- that's a bonus!). In this strain, I've already detailed at length how we need half of the planet dedicated as animal habitat in the space of a few decades. The Half Earth Project. Ambitious goals like this are a must, or as a person with poor astronomical knowledge once said: "Shoot for the moon. Even if you miss, you'll land among the stars." Hmph.

L is for legislation, meaning recognizing the rights of nature under law. Better people than I have written whole books on this subject, however the premise behind it is simple. An elephant's trunk doesn't make it luggage, despite what a dozen bad jokes will tell you. Animal personhood is the first tangible step towards societal reform. I don't expect to see elephants flying in first class, but neither should they be abused or held captive in poor conditions. Show some Humanners!

There's an argument against pretty much every intervention in nature which is that humans are 'playing God.' The implication of this is that we're crossing some line in the sand as to how much we're altering ecosystems. For better or for worse (more usually a bit of both), we've been 'playing God' for millennia. In other words, it's a bit late to stop now! Acting <u>like</u> gods is a different matter altogether, and something that occurs continuously across society. Stopping that is far more important, and requires only a bit of good old human humility.

The lesson to be learnt is one of perspective. Change is natural. Rapid, uncontrollable change less so. We should think of things not as 'good' or 'bad,' but as functions of their change- i.e 'worsening' or 'getter better.' Legislation is like changing gears on a bicycle: it needs to be well-timed and future-oriented. No-one wins by looking back at the hills they've already climbed.

E is for education, which I can tell you first hand is a flawed system. That's not an accusation, but rather an opportunity. With employment in a state of flux, we're becoming aware of the need for soft skills instead of hard knowledge for success. Communication, leadership, flexibility, conflict resolution, selflessness and grit are just a few of these. As great as string theory and the secundaria lingua of Latin might be, chances are they won't help you or I in saving the world.

The link to conservation is clear. With more capable, compassionate and confident youth, our environmental paradigms can be readily shifted. Ideas like Human Nature, Humanners, and Intelligence 2.0 provide us with landmarks to strive for. Conveniently, they also fit right in with the soft skills everyone's in favour of. So we need to teach these nuts and bolts of success, plus natural history as well. Kids should be learning how to learn and how to lead. As should everyone else.

A is for advancement, which means redefining progress to be nature-inclusive. At the moment, we're averting the blame for destructive development through biodiversity offset schemes. Habitat destroyed is 'compensated' by token input into conservation elsewhere. Remember rule 2: not all protected areas are created equal. Exchanging animal habitats just doesn't work.

What if we could link each country's currency with the state of its environment? So for example, removing old-growth forest in the UK would cause the value of the pound to drop. We'd get a new eco-market where nations would be forced to restore their natural heritage to thrive on a global scale. Of course, initiating such a system would require a near-inconceivable act of political courage and goodwill. But the change would set us on course for full ecological restoration within our lifetimes.

For advancement to take hold, we also need to change our language. 'Conservation' cannot mean rewinding the clock on our ecosystems. Nor can 'nature' continue to refer to pristine wilderness. We ought to approach the former of these as we do humanitarian causes. When the UN establishes a new refugee camp, they don't revert to using snail mail, candles and horse-drawn carts in honour of the 1800s. So why do we try to return nature to being as it was then, or even earlier? As for the term 'nature,' a similar logic applies. Just as they say your home is where your heart is, nature is wherever animals can be found. Including in the heart of the busiest megacities on Earth. Cities are humans' shoddy attempts at making ecosystems.

D is for dedication, at all levels. We all like to point fingers at the government or at big TNCs for our current environmental catastrophe. Unfortunately, they alone cannot fix it. Have you ever watched a flock of geese in flight? They (and many other birds) fly in an arrow-shaped 'v' formation. Each goose aside from the leader rides in the wake of the goose ahead of it. That's how we need to approach changemaking.

Continue the analogy and we arrive at a model I call "mass individualism." Just as the geese take turns to lead, we get places when individuals choose to act in synchrony, streamlining each other's efforts. Globalization has allowed us to connect in seconds to folks on the far side of the world. Give that technology to a Roman centurion and they would be dumbfounded. Had the Roman legions used cell phones, they would have been near-invincible. So let's stop taking it for granted and do something with it!

The 4, you may already have guessed, stands for Conservation's 4C's, which you'll remember from chapter 3. Connection, curiosity, creativity and collaboration are all necessary for any of this book's ideas to take hold. And as for 'CHANGE,' that's precisely what we'll achieve if we follow everything else up to that point in the acronym. And that's it: I PLEAD 4 CHANGE, your key card to the Human Nature club. Which concludes all 75,000 words of this book being distilled into a few short paragraphs. I bet you wish you'd known about that from the start!

One of my favourite quotes comes from Abraham Lincoln, who is credited as saying "The best way to predict your future is to create it." So far, we've pondered the past and pored over the present face of our human-nature relationship. We've only briefly considered the future. What does the path ahead have in store? With Lincoln's words in mind, let me set out my forecast for people and animals in the 21st century.

By 2100, the human population, though large, will have stabilized and begun a gradual descent. Contrary to alarmist prophecies, the human race will not be extinct. Apocalyptic events have been predicted since humans could speak. Anyone can make such a statement, though it's embarrassing if you live to see it proven wrong. Many Christian scholars and even the pope at the time foretold the end of the world in 1000 AD. Riots ensued in the lead-up to the new year, but when the prediction proved wrong the church simply rescheduled judgment day to 1033.

On 1st February 1524, twenty thousand Londoners left their homes and fled for higher ground. A group of prominent astrologers had predicted a great flood on that date sweeping across the world and cleansing it in Biblical fashion. Suffice to say, the waters were calm. The great explorer Christopher Columbus pinned the end of the world variously at 1656 and 1658, both of which dates he was 200 years too early to see. And in modern times, Jehovah's Witnesses used to preach of Armageddon taking place by 1975.

With this in mind, I think it's safe to say that humankind is more resilient than we tend to give ourselves credit for. We've spelled out our own death time and again, yet we're still very much alive and kicking. More reliable sources suggest we're in for a nearby supernova explosion and a large asteroid impact within the next few hundred thousand years. We're also likely to have a massive supervolcano eruption within a million years.

Taking place today, any one of these would spell serious trouble, but factoring in technological advancement, I reckon we'll make it through them when they come.

In short, I'm betting that *Homo sapiens* will last a good while longer. There's some very clever science and statistics behind the so-called Doomsday argument. This places human extinction in between 5,100 years' time and 7.8 million years' time at a 95% probability. That's also about the range we'd expect for a mammal species on Earth. We've got some solid estimates that the number of humans that will ever live will be 1.2 trillion give or take. We're many years away from reaching that total.

That's good news, right? We've got a long future ahead of us. The caveat is that if we screw up the Earth now, that span is drastically shortened. Intelligence works both ways now, and being foolish as a species means that you forego that long and prosperous future. I'm hopeful we'll set things right, and I've got two stories that might help to explain why. Firstly, whaling.

The industry of hunting down whales commercially arose in the 17th century, driven mostly by the demand for whale oil in candles. As this was phased out of use, other markets arose to take its place. Fertilizer could be made from whale bones, lubricants, soap and margarine from the body fat, and dog food from the vitamin C-rich meat.

It was war that saved the whales, inadvertently. Or the cold war at least. Rising hostilities between the US and modern-day Russia led to the American military installing underwater microphones to detect the noise of Russian submarines. That went well enough, but what they also found on the recordings were the previously-unheard songs of whales.

A biologist, Roger Payne, got hold of the recordings when they were declassified in the 1960s. A selection of them he released in an album titled *Songs of the Humpback Whale.* It was a hit- 100,000 copies sold, ranking it amongst some of the greatest musical hits of the time. Whale song popped up all over popular culture, and by 1972 a moratorium on commercial whaling was in place. That's just two years after the initial album release. It's incredible to think that Payne's half-hour tape catalysed such action, and a tribute to the speed at which public advocates can mobilize. Over 80% of all whales were killed in the first three-quarters of the 20th century. The whale song couldn't have come too soon.

The second story I'll share is about the dreaded CFCs. Chlorofluorocarbons (CFCs for short) arose in the 1930s as some very handy manmade chemicals. They met needs in aerosol spray cans and refrigerators with useful properties unmatched by any alternative. What we didn't know at the time was the damage they caused to the ozone layer in our atmosphere.

The ozone layer lies two dozen kilometres above the Earth's surface and acts as a sunscreen coat for the planet. It blocks out 98% of harmful UV radiation from the sun. This is the stuff that causes everything from sunburn to skin cancer and cataracts. In 1982, a team working at the South Pole noticed a hole in the ozone layer directly above Antarctica. That set alarm bells ringing, and blame soon fell on CFCs as the cause. In as little as five years, world leaders put in place legislation to phase out and shortly ban the chemicals. By 2065, it's predicted that the hole will have healed completely.

So environmental policy will be stronger in 2100, and better enforced I hope. We'll have at least 50% of the planet as safe space for animals, and hundreds of new national parks.

That's not bad given that the first national park was made only 150 years ago. Recently, rivers across multiple continents have been granted legal personhood, so why not animals like chimpanzees? I'm betting that most countries will end up recognizing the rights of nature at a macro scale. They'll bestow such privileges upon ecosystems or major landforms. What would excite me far more is if they were to continue the trend to the micro scale, benefitting individual animals. Like Switzerland today whose constitution promises that the dignity of all living beings will be upheld.

I already confided back in chapter 6 that I'm not a plant person. But I'm convinced that our expanding wealth of knowledge on non-animal groups like plants, bacteria and fungi will have profoundly changed our interactions with them. We now know that mother trees recognize and favour their offspring, making room for them, warning of threats, and sending nutrients through their root systems. Plants of all species exchange information and resources underground in what's known as the wood wide web. Birch trees can recall specific events that happened four years ago, and so learn from every challenge they overcome. Surely species that have developed their own internet deserve a little respect? The wood wide web improves their survival by a factor of four, which is more than can be said for our crummy human equivalent. And in my opinion, anything that makes it to 5000 years of age deserves a good pat on the bark!

Slime mould is the subject of another recent scientific shock. The stuff looks like moving threads of furry marmalade, but it can solve the most complex of mazes that you or I would struggle with. It has been presented with real-life transport problems and arrived spontaneously at a map of Tokyo's rail system and of the UK's major highways. Did I mention it doesn't have a brain?

Apparently, you don't need a brain to learn, to have a memory, or to problem-solve. And that's not it, because there's a type of slime mould that forms its own slimy civilizations. The mould spends most of its life as hundreds of thousands of tiny free-swimming creatures. At certain intervals, however, one will send out a signal that causes them all to gather into a large, slug-like form. Somehow they coordinate their actions so perfectly that they are able to operate as a single body. The 'slug' then transforms into a long stalk with a spore-emitting ball on top. All the slime mould creatures making up the stalk sacrifice themselves to allow those on top to reproduce: insects carry off the spores to start off new generations and the remarkable life-story repeats.

What constitutes brainpower will certainly need a good rethink as a result, and perhaps we'll learn a little humility along the way. After all, it is a little embarrassing to be beaten by a brainless lump of furry marmalade. What I find most fascinating about this is how our human landscapes will change from a major perspective shift. With animals as valued members of society, we might inhabit residences more akin to Kenya's 'giraffe manor.' Built in 1932 and modelled off a Scottish hunting lodge, this tall, ivy-covered building has become world-famous on account of its peculiar guests. In the 1970s, populations of Rothschild's giraffes were plummeting, and the manor's grounds became sanctuary to a small remnant herd. The animals were and still are left free to roam, but return to the building for feeds. Their long necks craning through top-story windows are a highlight during breakfast at the hotel today, delighting an array of international guests.

If you're anything like me, you're probably thinking that accommodation couldn't get any better than that. But staying there on holiday and owning the place are two different things. I don't expect most households to be welcoming giraffes to dinner any time soon. What I do expect is that we'll have changed just enough to welcome a few more animals into our homes. There's a prison in Florida that houses an exotic zoo on its grounds. A giant tortoise, a sloth, a Komodo dragon, an emu, skunks, lemurs, snakes and alligators all get provided for by the inmates as part of their chores. They love it: many return once their sentence is over to shows the collection to their families, or as a volunteer to help care for the creatures. What if all prisons were like that, or all schools? There'll never be a shortage of animals in need of care.

By 2100, conservation too will have reimagined itself. Already there are plans for building a 'Noah's Ark' on the South African coast. The centre (not a real ark) will house examples of a huge array of species, breeding endangered animals to conserve them and housing everything in futuristic domes. These can be climate-controlled and terraformed to match closely with the ecosystem the animals inside came from. The project is of unprecedented scale and due to be complete within a decade, providing a failsafe against extinction for the species it will hold. It'll open as a high-tech zoo for the public, whilst housing frozen cells and seeds from a large fraction of all plant and animal species as a backup for conservation efforts.

In the field, the tools at our disposal will also improve over the coming decades. We're rapidly reaching a stage where we can catalogue all the species present in a rainforest by looking at the blood from leeches. It's a system similar to the eDNA I mentioned in chapter 2 allowing insight into marine ecosystems and our good friend Nessie. Once we master these techniques, we'll be able to monitor the fates of all animals much more closely, and might be able to stop silent extinctions before they can happen.

We're also improving our knowledge of the past: finding nifty ways of analysing thousand-year-old poo to tell how humans have changed environments. And if we know what damage we've caused, it's much easier to compensate and set things right again, using our amazing abilities as 'ecosystem engineers' for good.

There'll be big benefits for us by 2100: transparent, flexible batteries modelled off fish scales, improved contact lenses inspired by gecko eyes, and psychedelic, colour-and-texture-changing clothing based on octopus skin. We'll have stronger-than-steel supermaterials from spider silk and self-healing, stretchable phone screens courtesy of jellyfish. That's just the tip of the iceberg. Chances are we'll boast extra senses too. We might have super-sensitive whiskers or thought-controlled tails, exoskeletons that give us super-strength and improve hearing, touch and smell, or sight like a mantis shrimp that can see ultra-violet and polarized light alongside more colours than you can imagine. By 2100, there'll be a human colony on Mars, learning from animals back on Earth how to survive, how to regulate conditions for human life and cope with the challenges the new setting brings about.

Whether we will have arrived at a fully green economy or not is hard to say. What I am confident of is that big businesses will have gone to far greater lengths to ensure corporate social responsibility, assisting in the process. A tiny proportion- as little as 0.1%- of shareholders to large TNCs have the power to control 80% of the company's value. These are the people most invested in the ventures, with the most to lose if it fails to grow and/ meet expectations. There's a growing consensus that businesses should go green, which by the turn of the century will be an unspoken agreement. Companies that fail to give back find themselves under mounting consumer pressure to change, so we have a positive cycle taking place.

I'd like to think we'll be better animals by 2100. We've got a long way to go. Nature enshrines efficiency, whilst humans prefer it on an as-needed basis. The Fibonacci sequence is some of the most beautiful mathematics yet described. The sequence goes 1,1,2,3,5,8 and so on, each number the sum of the two preceding. It was first set out in one of the very first books to use our current number system centuries back, but ever since people have noticed it popping up in the natural world. If you draw a series of squares with Fibonacci number side-lengths, then the spiral connecting them is a shape common across shells, flower petals, hurricanes and galaxies. The pattern of branches coming from the trunk of a tree, the arrangement of scales on a pinecone, the shapely proportions of animal bodies, the ratio of male to female bees in a hive, and the vision of birds of prey all obey rules related to Fibonacci's magical numbers.

No, it's not a conspiracy theory. They're simply a sequence of numbers that produce amazing efficiency. And evolution loves efficiency because energy wasted is energy no longer being used for an animal's survival. Fibonacci's numbers are a tried and tested formula for success, which we ought to pay more attention to. Thanks nature for the tip!

Even the apparent opposite of efficiency- art & aesthetics- is a realm in which nature has insight to contribute. When a Swedish newspaper acquired a few paintings by a chimp from a local zoo and hung them in a gallery, no-one questioned their alibi: 'produced by the great avant-garde artist Pierre Brassau.' One critic said "Brassau paints with powerful strokes, but also with clear determination. His brush strokes twist with furious fastidiousness. Pierre is an artist who performs with the delicacy of a ballet dancer." And why not indeed? If human art is indistinguishable from that of animals, then there's no reason they can't gain recognition for their creative works.

When it comes to the big questions, nature gives us a fresh perspective. Trees grow faster as they age, whilst humans shrink and shrivel. Theoretically, a tree could live forever. A human being can't. So we squander whatever resources we can get our hands on. Did you know there's sand in your toothpaste? We're rapidly running out of this seemingly infinite resource, depleting it in natural settings to build out concrete worlds. The solution is to mimic nature's cycles. Animals browse on trees, stealing their nutrients; eventually, an animal dies nearby and some of its nutrients pass into the soil; fungi break down soil nutrients and deliver them to the tree. Apply that principle to business and we arrive at a circular economy.

Animals are our past, present, and future. The cod that Vikings ate provided them with the vitamin D they needed to live healthily in extreme Northern winters. It allowed them to discover America and invade Europe, instead of being crippled by disease. If it weren't for cod, history would be very different. Our reliance on nature will never change.

In Southern Brazil, local fishermen have worked with dolphins for centuries. The dolphins herd the fish ashore and benefit from their fishermen partners blocking the escape. A pincer movement is formed by these two parties, and both are able to reap the rewards. For thousands, perhaps millions of years, tribes in Southern Africa have worked with birds called honeyguides to locate beehives in tree cavities. The hunters smash open the hives and take the honey, leaving the wax and calorie-rich larvae for the honeyguides to feast on. With any luck, the future will see many more human-animal partnerships forged, and of even greater complexity. Nature is a wonderful ally.

Have you heard of the 'six degrees of separation' idea? The theory goes that any person on Earth is linked to any other person by a maximum of six others- friends, acquaintances, distant relations. It's hard to prove (or disprove) and relies on a lenient definition of what counts as a connection. But whilst the human population keeps on growing, the number of degrees of separation between us is becoming less. In nature, food webs of predator and prey, parasite and host, see that the same applies. We as humans can have a direct effect on any animal on this planet through a short string of ties. That might sound scary, but it means we're still connected enough to change things. Powerlessness is simply a lack of meaningful choice. And if this book shows anything, it's that we have all the options.

In the late 19th century, a Polish eye doctor invented a new language he called Esperanto. It was based on a mix of half a dozen major European tongues and hoped to become the universal world language of the future. It's not made it quite that far, but still, Esperanto ranks as the most-spoken constructed language and has two million dedicated speakers. What its inventor failed to recognize is that Earth already has a universal language spoken by hundreds of trillions of creatures. Nature's laws are common across all Earth's kingdoms, and provide us with a rulebook for saving the world. What better way to finish?

Rule number one is that great minds think apart. No two humans are identical, and the same is true of any species you care to name. When you meet a stranger on the street, you don't say "Hi, human!" or "Hi, person number 99273751!" Unless you're living in some dystopian world, you recognize that both you and I and every other human are individuals. It's entirely natural to apply the same logic to other lifeforms: one macaque will be different from the next macaque, each ant from the rest of its colony. These differences provide the fuel on which evolution runs– small variations that slowly grown apart.

Science and society view species as they do shampoo or skincare products: each one a brand label promising certain features. Each label encompasses many bottles, every one of them identical. But animals are not clones! Personality and self-awareness, heightened emotions and consciousness all can be observed in the animal kingdom, yet still we try to reserve these merits for ourselves. Once we look beyond the smoke-screen of generalizations, false truths and sweeping conclusions new realms of thought appear. We see Humanners and Human Nature coming to the fore. But we also begin to see reflections of ourselves: an alpha chimp's dominance displays in a politician's blustering speech, or a young child's curiosity in an elephant calf just learning to use its trunk.

Importantly, we now attribute fresh value to our animal cousins, because conservation should not be a selfish exercise. We hear of giant pandas being bred and released to restore China's national pride, or so that our grandchildren perhaps might see them. Rather we do so as recompense for the damage already wrought on their kind, as a feeble kind of apology and a making amends.

Rule 2 is that big change comes in small packages. In print, the human genome would require as much space as in 5000 copies of Darwin's "On the Origin of Species" to write out. Some would call it the greatest wonder of the natural world, and yet all of it is contained in DNA strands so small as to be invisible to the naked eye. Where random variation between animals is the fuel, DNA is the raw materials and factory combined that forms the very first step for evolution to proceed.

You've probably heard of the IUCN Red List. It's a mammoth database of over one hundred thousand species, assessing exactly how close each one of them is to extinction. A pessimistic outlook for a dying world. A rather more recent concept is the IUCN Green List: a project with a similar intended scope, but one crucial difference. Instead of being some sort of doomsday clock for life on Earth, this flips the image and focuses only on those species that are recovering, experiencing marked success. It might seem a feeble gesture in the face of widespread declines, but this movement of 'Conservation Optimism' is truly a force to be reckoned with.

Put simply, inaction is the bane of conservation. So by recognizing and feting these smaller milestones, we gain much-needed momentum, boosting morale and fighting against apathy at large. Youth are a lever of sorts, able to see the bigger picture- and Greta Thunberg with her School Strike 4 Climate movement attests to the power each and every one of us holds in this delicate balance of paths.

Rule 3 is that everything is history. Nothing is new under the sun, and change has always been constant. Evolution, just like history, has a habit of repeating itself. Adaptions that work well are seen over and over again across species separated by vast tracts of time. These designs are the blueprints for success, so to speak, because there's no such thing as plagiarism in "survival of the fittest." What about us? Well, believe it or not, there are some Einsteins amongst the sheep who've clocked that 4 billion years testing ain't bad for product reviews. A new field of biomimicry has grown, which we've encountered throughout this book in an array of nature-inspired inventions. Fish cars, spiderweb glass, and firefly lightbulbs are just a few of the many treats we've seen arise from biomimicry.

Nature-inspired practices work wonders for environmental management too. We're beginning to recognize the importance of ecosystem connectivity, building new circular economies, employing systems thinking and holistic practices across the board. With these techniques at hand, we can stop acting like some malevolent God figure and more as a fairy godmother in nurturing nature back to health.

Rule 4 is that patience will hurt you. The thing about evolution is, it can be slow, but once it starts to make a move: once a species begins to diverge, grow a limb or a funky head-dress, it all takes place in the blink of an eye. Scientists have made snails turn slug in a single generation, their shells disappearing from their backs. Some traits can be picked up by a creature in its lifetime and passed on to its offspring, completely bypassing the normal tedious route.

The key lies in selective pressures: new variables introduced into the circle of life, forcing species to change. Humans are one, indeed chief amongst them. But now we're at a stage where the main selective pressures we face are those we have imposed upon ourselves. Global inequality, famine and conflict, climate change, biodiversity loss- these are the problems we are so reluctant to address, because to do so is to confront our own human vanity. Nobody's perfect.

Stagnation brings with it a slow demise for any species stretched beyond its limits. We need to keep on evolving, striving to be better. People are good at that. Martin Luther King Jr was editing his "I Have a Dream" speech until the moment he got on stage, and then scrapped the script to deliver the most famous words of the century. Leonardo Da Vinci took 15 years to paint the Mona Lisa, antagonizing his sponsors but resulting in a masterpiece. I've done the hard yards in writing this book, building off countless torchbearers before me. Which means it's time to activate our leanings.

Forecasting is difficult, and my predictions for 2100 may very well not come true. It's like predicting the weather two months from today (sunny, with a high of 28 degrees). So I need you the reader to make it all happen. Remember Lincoln's words of wisdom about crafting the future, go forth, and share Human Nature with the world. We've just got one more rule to explore...

Practice doesn't make perfect. Evolution had billions of years to master its craft, but still humans came about through a series of mistakes. In a book about animals, it seems only fitting to end with some words on the human creature. We are a product of our environment, shaping it and taking from it as the opportunity arises. We might be better at making tools and toys than other animals, but we're a lot worse than a flamingo at standing on one leg! One of these takes dextrous hands and the other takes clever body design. Are they really all that different? We've needed ferrets to fix our airplanes and wasps to think up paper. Eels invented remote control, Mongolian mountain voles make road signs, and bonobos were first to the idea of picnics. We're not left with much for humans to lay claim to.

4-inch-long tadpole shrimps have lasted through the past 300 million years, including the worst mass extinction event of all time and that which put an end to the dinosaurs. If I were to give credit where it's due, I think I'd be putting the label of animal genius on them as opposed to ourselves.

Life is one great big video game. Right now we've barely made it past level 2 or 3. That's an advantage because it means we're able to change tactics and upgrade our perception of the world and our place in it. Hopefully, this book has helped in that step. I have been luckier than most people, and my not dying at the paws of a leopard has brought me to places I never could have imagined. Now I'm giving back.

When we look at nature, we see ourselves. As the saying goes, we can run but we cannot hide from the impact every one of us has on this planet. Saving the world is fun, I guarantee it. So why not give it a try?

Afterword

I believe it's custom at this point to explain away my years of toil that went into this work. Then again, custom's never been my forte. In actual fact, this piece took shape over a matter of months during the dull spell that was Sydney's coronavirus lockdown. Many of the musings, examples and anecdotes stretch back much further, but I'm an avid collector and such things rarely go unshelved in my life.

Far more pressing than this self-indulgent chit-chat is the need to acknowledge those who've played a part in this merry dance of mine. First, to a leopard that spared my life- few things provide more food for thought than a near-death experience. Next to my family and all the other animals that have joined it over the years: to my mother for letting me raise assassin bugs in the living room; to my father for teaching me to rescue pelicans; to my brother for stomaching my dreadful puns and to my sister for scorning them.

I'd like to pay tribute to all those who helped launch me on this journey- far too many to name. To Megan & Naomi from Intrepid Landcare, to the kind-hearted Steve Parish, to Aurelie and Birdlife France, to Crista from CoalitionWILD, to Graham at WildEarth. And to all my old school friends for laughing just enough at my ideas. My thanks to Jane Goodall for her encouragement and sage words, to the team at *No Such Thing as a Fish* for expanding my horizons, and to the likes of Gerald

Durrell and David Attenborough for their constant inspiration. Also to Rituraj for talking of falling birds and other things, to Hein at the Natural History Museum for her knowledge of pigeon skins, and to the countless scientists whose thought-provoking work blesses the pages preceding. Thank you.

My sincerest gratitude goes out to my colleagues at Human Nature Projects: Justice Senkoto, Ronald Kaboye, Mbarek, Dolapo, Dmitriy, Harold, Yannick and many more. You prove to me that anything is possible, changing my life and inspiring me day in, day out. Without you this vision never would have come true.

Last of all, I'd like to thank you the reader. If there's a section that nobody reads, it's the afterword. Kudos to you for making it this far through. Honestly, I'm a fresh-faced 18-year-old with my life stretching out ahead of me. The next chapter of this great unfolding narrative comes from people like yourself taking up the message and sharing your thoughts. Reach out-let me know what you think of Human Nature, where you agree and disagree, or why I'm mad. I promise I'll read it all. That's it for now, so I'll say 'bon voyage' for whatever lies ahead, and remember to be a better animal.

Animal Cheat Sheet

If you have come across a suspicious critter between the front cover and this, chances are there's a description here to put you at ease. Happy referencing!

...

Acacia: a tree of the African savannah famous for its inch-long thorns, added by dictionary writers alongside 'aardvark' to spice up the start of their books.

Albatross: think seagull and enlarge by a factor of five. Often described by lost sailors as 'wandering' to lessen their fears and improve morale. This bird is never lost.

Algae: a generic term to describe green slimy stuff found in water. This makes a rather unsavoury but nonetheless helpful lodger for corals.

Alligator: an unsuccessful attempt of nature to forge political goodwill between nations, the two species sharing this name are crocodilian diplomats to China and America.

Ant: a possible model of the future human race, once we've perfected biological shrinking and learnt to cooperate. The most successful, and humble, of all the insects.

Armadillo: the good cop/bad cop routine of the animal world. There are tank-like banded armadillos (Mk. 3,7 & 9) and pink fairy armadillos that wouldn't hurt a fly.

Ass: hardly a term of endearment, the name applies equally to a donkey-like creature, a stupid person, and one's rear end.

This animal definitely got the bum deal.

Bacteria: occasionally known by the less endearing name of 'germs,' these are the invisible creatures that run the world. Sadly, the Hitlers of the bacteria realm give them all a bad name.

Bat: a poetical creature, blessed with a name to which almost infinite rhymes exist. Not to be confused with the sporting implement, this is an almost-bird.

Bear: cousin to bigfoot, but less camera-shy. 'Grizzly' describes those who've not recently shaved, and 'polar' those with divisive opinions. Every good bear deserves fish is the phrase they're taught at music school. Having failed at busking, they raid cars to get it.

Beaver: near-unparalleled in personal branding, the beaver has gained a reputation for industry despite spending all day asleep. Its anal secretions were once used to flavour vanilla ice cream.

Bee: immortalized in Shakespeare's "to bee or not to bee," these are the ultimate workaholics, and so live on a permanent sugar high. Males bees are called 'drones.' That term also describes most male humans I know.

Beetle: armoured vehicles to the last, most take on a professional black tone but a few rebels opt for fluorescent hues. The bombardier beetle wins on the naming front, joining the elitist ranks of beetle military with a powerful arsenal to match.

Bison: looks rather like a bull with a half-body afro, only don't say that to its face! Short of temper and quick to judge, the wrong side of this animal is definitely right in front.

Bonobo: a chimpanzee that's failed to find a cure for hair-loss, still taking late-night trips to the shops to try out the latest miracle products. Laughing, kind, chatty, this is everyone's favourite customer.

Bowerbird: hell-bent from birth on an artistic career, this bird is better suited to architecture. A hoarded and pilferer, the bowerbird is a museum's worst nightmare.

Buffalo: the bison's uncle, lacking the afro but hiding the fact with a thick layer of mud. The buffalo has an impressive set of horns but still looks stupid in photos.

Bush-baby: an unskilled taxidermist was working on a mouse-lemur whilst drunk at 3am in the morning, maybe blindfolded. The glass eyes selected for it are at least seven sizes too large. In animal literature, it serves as Frankenstein's monster.

Butterfly: with a name coined in a cookbook, this animal's title is also conveniently onomatopoeic for its only encounters with man occur when a Scottish bloke finds one indoors- "Awa' yer wee flitter; our the windae and ye'd better flie!"

Capuchin: movie-star monkeys, capuchins share screen time in *Indiana Jones*, *Night at the Museum*, and *Pirates of the Caribbean*. The last of these was a grudging role, having failed the audition for capuchin Jack Sparrow.

Capybara: a rat-pig the size and shape of a small whisky-keg, with the vacant expression of a kid at the back of a chemistry class on Friday afternoon.

Cassowary: the subject of a peculiar creation myth in which a runty male ostrich ran headfirst into a freshly-painted door. Very self-conscious.

Cat: a lion with growth problems engaged in a conspiracy with dogs to break the internet. Also the cold-hearted (bird-)murderer you keep in your living-room.

Chaffinch: a salmon-coloured bird just larger than a golf ball and much less painful in the back of one's head. Wild chaffinches eat seeds from bird-feeders, whereas wild golf balls eat brain cells.

Cheetah: a desperate loser, small enough to be picked on by lions, leopards, hyenas, wild dogs and vultures. Despite the odds, this is a very honest creature.

Chicken: a feathered fowl unique in waging a philosophical war against that which comes out of its rear end- the egg. Like the cheetah, chickens suffer from anxiety problems on account of negative stereotypes.

Chimpanzee: lacking a charming smile, chimps have broken into Hollywood as badass fighters, in *Planet of the Apes*. Ironically, they use guerilla tactics.

Cobra: if, like myself, you speak fluent snake, then you'll know that all this creature ever says is "piss off." You'd be well advised to listen. Still, some idiots put them in baskets and play them music they cannot hear. Charming.

Cockroach: the only animal for which 'killer' tops google's auto-complete. Their ability to withstand forces 1000 times their body weight might explain why. Foley artists use them to create realistic scream sounds when dropped down the neck of an unsuspecting victim.

Cod: battered or otherwise, cod are hardly the most attractive fish. Having had a glass of straight lemon juice and applied a hasty coat of red lipstick, you might look similar.

Coelacanth: kind of cute in a grotesque, deep-sea way. Think of a fish-shaped balloon with too many fins, tiny baby teeth and a paint job inspired by Van Gogh's *Starry Night*.

Condor: big, black, ugly vulture. The South American sort has improved itself with an Elizabethan ruff. The North American variety looks like it's had its head boiled.

Coral: ocean animals in the guise of alien plants. Brain corals have the largest brain-to-body-size ratio of any animal. Sadly scientists think they're a bit thick.

Crab: drunken boxers to the last, crabs can't walk straight and suffer from vision impairment. A few, the hermit crabs, have opted out of this hedonistic lifestyle.

Crayfish: though generally better off than crabs, both animals are threatened by their natural predator- the Cajun chef. Crayfish are not fish because fish do not have claws.

Crocodile: alligators gone rogue, crocodiles are wily creatures. They allow small birds to pick their teeth as if inviting you to try. I don't recommend it.

Crow: crows and ravens are the subjects of a tragic case of mistaken identity. Mortal enemies, most people use these two terms indiscriminately. Ravens are the big bullies, with hoarse calls, nasty-looking beaks and a love of old castles. Crows are pacifists.

Cuttlefish: the only underwater creature to do a convincing zebra impression. Just as octopuses are not cats, cuttlefish are not fish. Don't argue.

Deer: an old-fashioned animal reserved for written letters, the deer sadly lacks much of a following in the modern world. Chinese water deer compensate by growing fangs.

Demon Duck of Doom: an ancient bird Godzilla, artists somehow always manage to make the demon duck look happy with a beak that could bite off your head.

Diplodocus: sporting a neck to make a giraffe cry and a tail of foolish proportions, the diplodocus clearly had only one goal in life- to be remembered as the largest animal of all time. Sadly dippy the dinosaur was pipped by the blue whale on this, and so went extinct.

Dodo: famous for dying, and for its possession of all the stereotypical French characteristics. It forms fine cuisine, is partial to bread and wine, often rude and smelly, looks better in a wig, has a superior attitude, a faux bravado, and... okay, I'll stop there.

Dog: man's best friend until cats and computers arrived on the scene. Dogs have since been relegated to the halls of Buckingham Palace and several other habitat fragments.

Dolphin: hides its financial troubles from exorbitant dentist bills behind a faux smile and inane chatter. To add insult to injury, it doesn't even get cast on toothpaste adverts.

Dove: a pigeon with manners, yet hated by city-dwellers globally. The rock dove is our common-or-garden rat with wings; the turtle-dove is just plain confused.

Duck: an animal loved by spectators of swing tennis and boomerang-throwing, who frequently shout its name. If on land, the collective noun is a 'badling,' whereas in water a 'paddling' is the correct term. If one foot only is in the water, ask someone else.

Earwig: take a slightly squashed cricket and insert a pair of tweezers up its rear end. This insect should not be confused with the niche fashion craze amongst 18th-century snobs.

Echidna: the bloated gipsy cousin of the hedgehog, it has a nose to make Pinocchio proud but gets rather antsy if anyone makes such a comment out loud.

Eel: nature's chief knucklehead, it seeks to regain its marbles by gawping like a circus laughing clown. Nobody has ever seen one breed and consequently eel porn is all that biologists dream about.

Elephant: a majestic African giant, the elephant is in a permanent state of embarrassment on account of having lent its name to all sorts of poor companions. The rat-like elephant shrew is tiny. Elephant seals with their half-trunk just look silly.

Elk: one of the more dashing deer varieties, elks love cold weather that makes their breath fog up. So it is ironic that their heads frequently adorn blazing fireplaces.

Emu: a cassowary that went in with the laundry to wash off its paint. Famous for having defeated the Australian army in the 1932 Great Emu War. Not worth crossing.

Ferret: a relative of the weasel with a claim as the world's first animal electrician, carrying cables. Personally, I think they're better suited to a gymnastics career. Surely they'd pre-ferret.

Fish: Doesn't exist. Look someplace else. This one's a red herring.

Flamingo: although related to pigeons, flamingos keep themselves in our good books by producing pink milk. They are adapted for yoga and can balance on one leg even when dead.

Flatworm: does what it says on the tin i.e a worm that is flat. Would have potential as new-age flying carpets if it weren't for their size. And their inability to fly.

Fly: mocks flatworms relentlessly. Most tell it to buzz off. Its infamy in the insect world is increased by its inclusion in 'waiter, waiter...' jokes.

Fox: best known for jumping over lazy dogs during typing lessons. Very few animals have an 'x' in their name, and oxen aren't all that good at jumping. Where the sentence falls down is in the fact that foxes aren't brown. Or maybe we're missing the backstory?

Frog: hardly discerning creatures, frogs are the only group of animals that are aroused by both canon-fire and washing machines. Not often simultaneously.

Fungi: aside from providing real-estate for fairies and sprites, fungi live rather unglamorous lives. Plants overshadow them, and they're altogether mistreeted.

Galapagos tortoise: an inspiration for Darwin's theory of evolution and a darn good meal too. Usually described as 'giant,' which must be some consolation.

Gecko: a paradox, the gecko is smart enough to dine off insects attracted to our ceiling lights yet dull enough to have no concept of gravity, or else it would fall.

Giraffe: having a selfie stick for a neck means the view is to die for, but of course scuffles are inevitable. Fighting like a giant Newton's cradle, a pair of brawling male giraffes would make an epic but cumbersome desk toy.

Goose: a big duck, less commonly served with pancakes.

Gorilla: the bodybuilder ape, unusual in that it takes great pride in grey hairs. The look of a gorilla is that of your boss the day you're three hours late into work.

Hedgehog: see echidna, only cuter. A sidekick to cats in the animal conspiracy to break the internet.

Heron: the zen bird, capable of staring at water for hours until a fish swims by. Its appearance is similar to that of a private school Latin teacher.

Hippo: two bathtubs stuck together to create a submersible with a propeller that shoots dung. It may spend more time than any other animal yawning. Who wouldn't with teeth like that?

Honeyguide: a bird with a sweet tooth despite having no teeth at all. The honeyguide is a secret agent working to undermine bee colonies, with a secret whistle for backup.

Hyena: would-be king of the beasts, except royalty aren't allowed a sense of humour. Laughs incessantly like a Hollywood psychopath. Don't joke about this one.

Ibex: affix a pair of elephant tusks to the head of a goat and place it on the nearest rocky precipice. You'd think that having horns so large would hinder mobility, but this is one agile creature.

Iguana: the ultimate in tropical island lifestyles, basking eyes closed in a sunny patch with little ambition in life. Iguanas have one downfall, which is their excessive makeup.

Iguanadon: giant dinosaur iguana... sort of. Think of an old, stooped t-rex with a cow head and a turtle beak. You're on the right track.

Jaguar: big, spotty housecat. Place it in the Amazon stalking a capybara for added realism. Not to be confused with the luxury car brand, which is easier to spot.

Jellyfish: a squidgy lump with tassels. Doesn't do a lot except float and feed. Some are immortal. None are even close to being fishes, because fishes have fins and do things.

Kakapo: a hobbit bird that missed the cut for *Lord of the Rings* and has instead taken up method acting as a tree. It's surprisingly good at this.

Kea: the evil step-cousin of the kakapo with a nasty habit of eating live sheep. Its cruel, hooked beak may have been God's early attempt at creating a bottle-opener.

Kestrel: a tiny bird of prey whose superpower is flying without moving. Not to be confused with a kite- look for the string.

Killer whale: with looks to kill, these animals deserve a better name. They're deadly enough as hunters, but certainly not whales. What they <u>are</u> is dolphins that kill and eat sharks. Why not 'sharkbusters'?! False killer whales are not killer whales. Obviously.

Kingfisher: a niche role to be sure, but castle moats do increase drowning risks for royalty. Before lifeguards were a thing, something had to fish them out.

Koala: living teddy bears, koalas serve a function similar to baubles on a Christmas tree- they look nice. Koalas never move or do anything remotely interesting. As I say, they're purely decorative.

Komodo dragon: another letdown, this animal doesn't breathe fire, possess any magical abilities, or have wings. Its teeth are invisible, but it does have venomous spit.

Krill: tiny sea creatures slightly larger than paperclips named after a Norwegian word for small fish. They look more like lobsters and have at least a dozen legs. They are not fish.

Lacewing: the result of a failed experiment to cross a praying mantis with a dragonfly. Young lacewings are called lions or wolves and are like very deadly caterpillars.

Leech: here's one animal you definitely don't want to google. The worst posting for a biologist intern is with leech re-searchers, who collect specimens on their bare legs.

Lemur: named after ghosts, they sure look like they've seen one. With bulging eyes and crazed expressions, lemurs are the mad scientists of the monkey world. Should we really be surprised that they rub toxic millipedes over their genitals?

Leopard: see jaguar, only more intelligent. Whereas jaguars wear their golden coats in the greenery of rainforests, leopards have worked out that camouflage works best in yellow, grassy savannahs.

Lion: see leopard, only easier to spot. There's a reason why lion tamers exist and tabby cat tamers don't. The reason is that lion tamers are mad, and mad people don't work in halves.

Lizard: lizards have one problem- they can't run and breathe.

Normally if you're fleeing from a large predator, you shout something helpful to your companions like "run faster!" Lizards can't. Still, some survive hurricanes and others can fly. A few are well-known, such as the British royal family.

Locust: like a large cricket, but with a big red button. This is a literal kill switch, turning them from harmless to apocalyptic. Usually solitary creatures, they occasionally get together and wipe out civilizations through famine.

Lorikeet: one day the divine creator took a day off, and a three-year-old stepped in with four colour pencils. Thus the lorikeet was born. Most of these birds feed on pollen and nectar from plants, however, some prefer people's fingers.

Macaque: a group of monkeys described by scientists looking up their rear end. Hence 'lion-tailed,' 'pig-tailed,' and 'stump-tailed' varieties. No wonder they look disgruntled.

Magpie: Eurasian magpies are elegant, benign birds with a smart black-and-white suit. Australian magpies are brutes, most closely related to the butcherbirds. Enough said.

Mammoth: an elephant on a bad hair day. One big fan was US President Thomas Jefferson. I reckon he was mostly just after a new wig, though.

Manatee: an old, obese dolphin just rammed headfirst into a sea wall. It seems to have never quite recovered its looks, brains or dignity.

Manta Ray: with their big, black wings and night-time forays, mantas are the bats of the ocean. Instead of turning into vampires, they're used by mermaids as hoverboards.

Mantis shrimp: it's neither a mantis nor a shrimp, but pray don't tell it that. The mantis shrimp can punch so fast that the water surrounding them reaches the temperature of the Sun's surface. Do not try this at home.

Meerkat: the soccer animal, whose pose inspired the wall formation. Rather unsportsmanlike, they are the most murderous mammals known to science.

Microbe: any living thing too small for you to see. This may include car keys.

Moose: the largest member of the deer family, moose (or is it mooses or meese?) have a nasty habit of walking onto roads in front of oncoming traffic. Brake or be broken.

Mosquito: not to be confused with the cocktail that shares its name. One is cool, the other chills the blood. One has a mint leaf garnish, the other you're meant to leave out with the garbage.

Mountweazel: an odd critter that is the favourite pet of lawyers. Mountweazels pop up everywhere yet are still able to throw the casual observer. See steinlaus.

Mouse: this animal is under threat due to the invention of the touchscreen. Some commentators have raised concerns about animal cruelty in its usage.

Muskrat: see beaver, minus the awesome tail.

Muskoxen: see bison, only with a worse hairdo.

Nightingale: what bird-watchers call an LBJ (little brown job): a forgettable face and unimpressive physique. It makes up by singing seductively outside of poets' windows.

Octopus: some scientists have claimed that octopuses are aliens, or mutated squids infected with DNA from an ancient meteorite. If this were true, we should have our global defense services on red alert: we have an alien incursion on Earth. And they're armed.

Orangutan: the one animal everyone wants to look like... or is that just me? No other creature could look so dignified covered in orange hair, hanging upside down from a tree.

Orca: see killer whale, except this name is marginally better. Scientifically, the animal is called <u>Orcinus</u> orca, which just sounds silly.

Otter: alone amongst animals in its ability to look adorable regardless of its age, the otter also swims a remarkable backstroke. You really otter be impressed!

Panda: a remarkable sight, if only because a black and white sofa has animated itself. Pandas have a privileged education, passing through panda preschool and eventually on to Harvard. I'd kill to follow their curriculum, instead of everything being black and white.

Penguin: the most horribly pretentious birds, who'd rather bathe than waste energy flying. Plenty of them adopt titles such as 'king' and 'emperor.' There are beefeater (chinstrap) penguins, but when a 'royal' penguin species was described others sought to discredit it. With great power comes great vanity.

Pigeon: an uncouth, demanding, worrisome bird like the family member you never invite round for dinner. Not to be confused with the well-mannered dove, distinguishable by its demeanour.

Platypus: appears between perplexing and problematic in the dictionary. It's not hard to see why. It has a duck's bill, an otter's body, a beaver's tail, venomous legs, no stomach, and lactates but lacks nipples. Having seen a few, I'm inclined to believe they're a hoax.

Porpoise: a younger dolphin swam into a wall. End of story.

Puffin: sharing both its stature and colour scheme with the queen, one might think this was a pretentious bird. Far from it, and indeed the puffin prefers rock outcrops to rotten old palaces. It would not dare to claim it owns all the swans in

Britain.

Pygmy raccoon: it looks like a bandit, smells like a bandit, and acts like a bandit. Its lock-picking skills have to be seen to be believed. In short, it's a criminal creature.

Rabbit: famous for carrying pocket watches and having very deep holes, rabbits have particularly busy schedules around Easter time when modelling for chocolate sculptures.

Rat: a scientist's best friend, although that sentiment is not reciprocated. The term 'lab rat' is strangely synonymous with 'guinea pig.'

Redstart: a robin going through a punk phase.

Rhino: looks permanently grumpy on account of the apartheid system it still suffers under. 'Black' and 'white' rhinos are both in fact grey, but the former is by far the worst off. Black rhinos live in rubbish habitats, eat rubbish food, and are more likely to get killed.

Robin: a small plum-pudding of a bird, often orange though mistakenly labelled as robin redbreast. Its preferred perch is a gardeners spade, and it looks damn fine in the snow.

Roundworm: one of a lucky few not to have found its way under a road train's wheels or a heavy boot. In both cases, the result is a flatworm.

Salamander: a frog-lizard, known to folklore as the creature that quenches fire, has a breath that makes humans explode, and that can only be killed by its poisoning itself. Unfortunately for the real creatures, they possess none of these practical attributes.

Scorpion: aquatic scorpions once ruled the seas. Nowadays they live in bitter resentment of the animals that overthrew them, relegated to the land and jabbing at any animal stupid enough to come near with that nasty sting of theirs.

Seal: born with an embarrassing, vacant expression, the seal evolved flippers to cover its face. Macho seals try to compensate with shark impressions and titles like 'sealion.'

Shark: a seal's worst nightmare, often accompanied by ominous music. Why they play this and so alert seals to their presence is a mystery. Most likely it's an advertising gimmick to get them in more films.

Shellfish: fish don't have shells. Full stop.

Skylark: see nightingale, only larger. Thus LBJ, in this case, stands for larger brown job.

Slime mould: before chefs worked out that mouldy cheese was delicious, there was an experimental phase with different foodstuffs. This was one of the failed attempts, leaving a pulped banana to putrefy.

Sloth: with Wolverine claws and Joker facepaint, the sloth is rather less intimidating than you might imagine. Most of the time they're just hanging around.

Slug: a small improvement on the slime mould experiment. A peeled banana left out long enough to go brown and squishy then brought to life. Alternatively, a homeless snail.

Snail: the most dedicated real estate agents, carrying a home around with them. Hermit crabs are their best customers; mind-controlling parasitic worms are their worst.

Snake: the pantomime villain, typically introduced with 'he's below you!' If you hear this, freeze. Snakes don't have legs, but enjoy watching people balance on one.

Song thrush: see skylark, with the brown paint job left half done. LBJ in this case stands for larger brownish jerk. The latter part owes to their habit of smashing open snails' shells.

Spider: most content themselves with sticking to the shadowy corners of ceilings. The less intelligent opt for bathtubs.

They don't need second hand stores because they start with eight.

Sponge: a cleaning implement. The more you pay for it, the greater the chance it will resemble the sea creature of this name.

Starfish: neither a star nor a fish. Everyone knows stars don't have five arms. And fish don't often vomit out their entire stomachs.

Starling: a shiny clean blackbird has just lost a paintball fight to an army of ants. Small wonder it looks cross and will eat whatever insects it can find.

Stork: a heron that has befallen a freak accident. Having suffered a scalping this bird has entirely lost its poise and its patience.

Sundew: an octopus-plant that looks as though it's just been through a heavy rainstorm. Its name is an oxymoron, just like the whale shark, mantis shrimp and chicken frog.

Swallow: these birds are everywhere, and so the only mystery about them that remains is their flight velocity. Any attempt to photograph one or to follow it with the naked eye results in a streaky blur and a strong urge to blink. A group of swallows is called a gulp.

Tadpole shrimp: think of a tiny pancake shaped like the starship Enterprise with a snake's tongue sticking out of its rear end. You're close.

Tardigrade: some know this as a water bear. If by 'bear' you mean cyborg mole with eight legs, then you're not far from the mark.

Tasmanian devil: imagine an ugly black dog lying on a football pitch when a striping machine passes and paints across its chest. Add the strongest relative bite of any animal.

Termite: termites look like ants with developmental issues. I guess the ants feel a bit miffed that termites still manage to build better mounds than them. A queen termite has several gooey marshmallows attached to her rear end.

Thylacine: a cross between a tiger and a shaved poodle. Anyone who can replicate this hybridization may claim a million-dollar reward for rediscovering the extinct species.

Tick: a spider zapped with a shrink ray and promptly ironed. It exacts revenge for this treatment by placing humans on its menu.

Tiger: a lion had several fake tans then got into a fight with the striping machine that ran over its tazzie devil friend. Its sole wish in life is to grow a mane.

Tit: poor, poor creature. Lumped with such a foul name, the tit is a paragon of innocence. It is a baby bird that never grows up, yet has exquisite manners in its own petite way.

Toad: the term used for frogs that no-one likes. A favourite of witches, who detect a certain warty resemblance. 'Bubble, bubble' means a toad's in trouble.

Tortoise: a turtle that made the mistake of evolving for life on land. It regrets it and has to live with the consequences: an ambling, awkward gait and tiddly little arms. Its shell is used for sulking.

T-rex: tyrannosaurs prefer meat to mates, so it's a wonder this creature has a nickname. Perhaps 'T-rex' is easier to shout when running away from the giant killer dinosaur.

Triceratops: genetically engineered for jousting, this boasts both shield and lances on its head-parts. Do not approach from the front.

Trilobite: take a large tongue and put it through a bread-slicing machine, then add a croissant as a head. You're probably nowhere near what this creature looks like, but then again, I'm not a chef.

Turtle: looks permanently like it's going to puke. Maybe that's why they never open their mouths? Those with stronger stomachs lead a double-life as ninja vigilantes.

Velociraptor: like in Jurassic Park, only feathery and half the size. Personally, I think that having a banshee bird-dino leaping at me would still be just as terrifying.

Virus: being half-dead has certain advantages in making viruses nigh-indestructible. They're like the weeping angels from Doctor Who, just a teeny bit smaller.

Vole: like the bat, this is an animal invented purely for poetry. Most people would call it a rat, but those rhymes are all taken up.

Wasp: well, if there's one thing I can say for them, it's that they've got their colour scheme sorted. Yellow and black looks good on anything. Honeybees stole the look as cheap, knock-off copies.

Whale: if you inflate a dolphin to ten times its normal size and age it several thousand years, it probably starts to look something like this. Whale poop is nature's attempt at mass-produced fertilizer. Whales were also the original opera singers.

Whale shark: hard to fit in a tin, but if it did, it would do just what it said on it. The whale shark is exactly halfway between its two namesakes, though fish unions have caused it to be classed amongst sharks.

Wild boar: the opposite of a tame pig. Alternatively, a term of endearment applicable to one's ex. Not to be confused with the wild bear. For ID, look closely at the size of the teeth.

Wildebeest: these antelope exist purely for the purpose of

Select List of World Animal Days

- **5th January**: Bird Day
- **21st January**: Squirrel Appreciation Day
- **2nd February**: Hedgehog Day & Wetlands Day
- **5th February**: Western Monarch Day
- **15th February**: Hippo Day
- **27th February**: Polar Bear Day
- **3rd Sat in February**: Whale Day & Pangolin Day
- **3rd March**: World Wildlife Day
- **14th March**: Moth-er Day & Butterfly Day
- **16th March**: Panda Day
- **20th March**: Frog Day & Sparrow Day
- **21st March**: Intl. Day of Forests
- **22nd March**: World Water Day & Seal Day
- **Last Wednesday in March**: Manatee Appreciation Day
- **2nd April**: Ferret Day
- **4th April**: Rat Day
- **7th April**: Beaver Day
- **14th April**: Dolphin Day
- **17th April**: Bat Appreciation Day
- **22nd April**: Earth Day
- **25th April**: World Penguin Day
- **27th April**: World Tapir Day
- **1st May**: Save the Rhino Day

- **3rd May**: Koala Day
- **20th May**: Bee Day
- **22nd May**: Intl. Day for Biological Diversity
- **23rd May**: Turtle Day
- **2nd Saturday in May**: World Migratory Bird Day
- **3rd Friday in May**: Endangered Species Day
- **Last Wednesday in May**: World Otter Day
- **1st June**: World Reef Day
- **5th June**: World Environment Day
- **8th June**: World Oceans Day
- **21st June**: Giraffe Day
- **14th July**: Shark Awareness Day & Chimpanzee Day
- **16th July**: Snake Day
- **29th July**: Tiger Day
- **10th August**: Lion Day
- **12th August**: Elephant Day
- **14th August**: Lizard Day
- **19th August**: Orangutan Day
- **20th August**: Mosquito Day
- **30th August**: Whale Shark Day
- **8th September**: Iguana Awareness Day
- **1st Saturday in September**: Vulture Day
- **3rd Saturday in September**: Red Panda Day
- **4th October**: World Animal Day
- **6th October**: Badger Day
- **18th October**: Okapi Day
- **20th October**: Sloth Day
- **21st October**: Reptile Awareness Day
- **3rd November**: Jellyfish Day
- **4th December**: Cheetah Day & Wildlife Conservation Day
- **10th December**: Intl. Animal Rights Day

- **11th December**: Intl. Mountain Day
- **14th December**: Monkey Day
- **27th December**: Visit the Zoo Day

Full Eco-IQ Quiz Results

1.) What percentage of forest cover has been lost in the Brazilian Amazon from deforestation over the past 50 years?

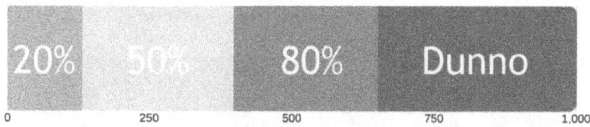

2.) Approximately how many living species are we estimated to share the Earth with?

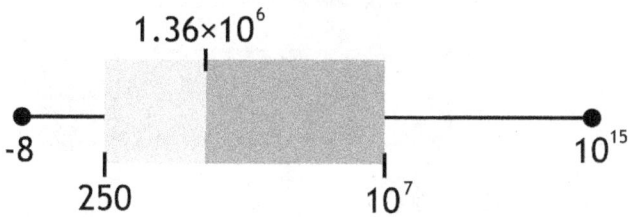

$$1.36 \times 10^{6}$$

-8 10^{15}

250   10^{7}

3.) How many of Earth's species has science currently described?

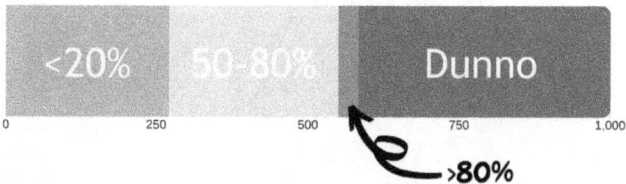

>80%

4.) How has the number of recorded poaching incidents of rhinos in South Africa changed over the past 5 years?

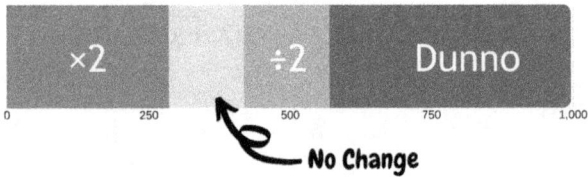

×2　÷2　Dunno

0　250　500　750　1,000

No Change

6.) Biodiversity loss is likely to impact progress towards approximately what proportion of the Sustainable Development Goals?

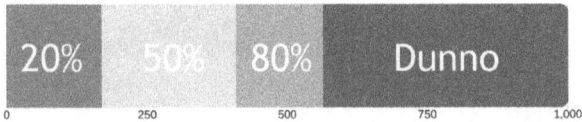

20%　50%　80%　Dunno

0　250　500　750　1,000

7.) Since 1970, the abundance of animals in the wild has...

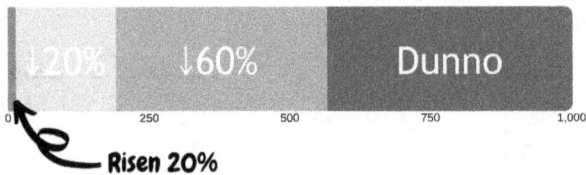

↓20%　↓60%　Dunno

0　250　500　750　1,000

Risen 20%

8.) What percentage of the world's economy is derived from ecosystem services?

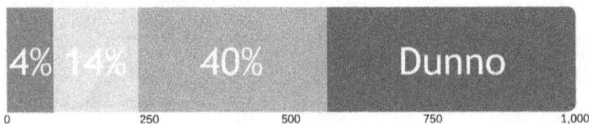

4%　14%　40%　Dunno

0　250　500　750　1,000

9.) If no major action is taken, how many degrees warmer will the Earth be in 2040 compared to pre-industrial levels?

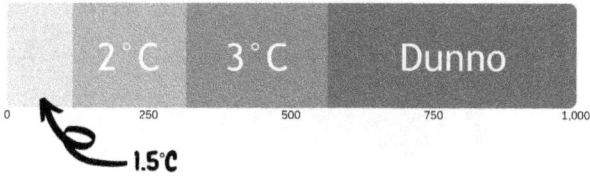

| 2°C | 3°C | Dunno |

0 — 250 — 500 — 750 — 1,000

1.5°C

10.) Do richer or poorer countries have more threatened animals?

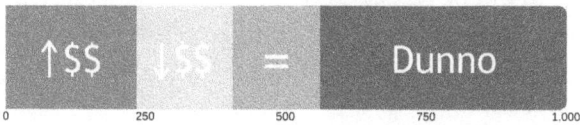

| ↑$$ | ↓$$ | = | Dunno |

0 — 250 — 500 — 750 — 1,000

11.) Threats scored on severity for terrestrial and freshwater ecosystems...

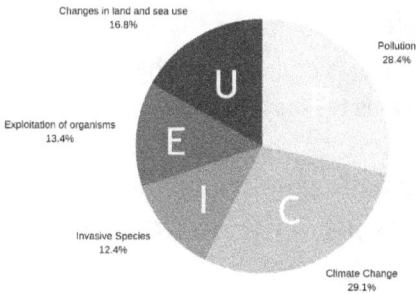

Changes in land and sea use 16.8%
Pollution 28.4%
Exploitation of organisms 13.4%
Invasive Species 12.4%
Climate Change 29.1%

Human Nature Crowdsourcing

Here's a shortlist of some of the most thought-provoking and popular answers to question twelve: "Provide one reason why humans might be considered as separate from animals." See what you make of them now, having read this book. Would you like to argue my points and continue the debate? Get in touch...

- "Humans haven't yet developed enough to see they are not."
- "Animals do what is instinctual; humans have choices and free will to choose to do good or harm in the world."
- "Animals can't speak and ask for what they want."
- "Evolution gave us fantasy."
- "Humans are idiots." Also "Stupidity."
- "Because we have the power to change things."
- "We are animals, just the most aggressive and arrogant species."
- "Humans could not survive without animals, but animals most certainly could survive without humans."
- "Humans are selfish. Animals are self-reliant."
- "We developed more powerful weapons."
- "Humans are the only species that destroy their own habitat."
- "Supposedly humans are more rational than other animals..."

jumping into crocodile-infested rivers during nature documentaries. Their front and rear ends bear a striking resemblance.

Wolf: sometimes found dressed in sheep's clothing, which can't be comfortable. Wolves create a bad name for themselves in fairy tales to prevent door-knockers and cold-calling.

Wrasse: with a name that's Celtic for 'hag,' these are some of the most attractive fish. Figure that. Many take up the profession of underwater washerwomen. Figure that too.

Yak: punk cows with hair that's constantly getting into their eyes. Who knew that dreadlocks looked better on animals?

Zebra: horses with foresight enough to evolve in premonition of urban planning. Perfectly camouflaged against street crossings, the sad result is that zebras face high numbers of pedestrian fatalities.

Zebrafish: copycats, if you'll excuse the mixed metaphor. Zebrafish aspire to live up to their equine heroes, but they got mixed up and placed their stripes the wrong way round.

About the Author

Elliot is the founder and CEO of Human Nature Projects, an environmental charity supporting volunteers across 105 countries. He is a TED speaker, popular science writer, podcast host and filmmaker with the goal of reframing our human relationship with nature. Living in Sydney, Australia, he spends his spare time caring for injured wildlife.

You can connect with me on:
- 🌐 https://www.elliotconnor.com
- 𝕏 https://twitter.com/eco_elliot
- f https://www.facebook.com/elliot.eco
- 🔗 https://www.instagram.com/elliotconnor.eco

www.ingramcontent.com/pod-product-compliance
Lightning Source LLC
Chambersburg PA
CBHW021856020426
42334CB00013B/361